General Botany

the text of this book is printed
on 100% recycled paper

About the Authors

Harry J. Fuller received the M.S. and Ph.D. degrees from Washington University and the Missouri Botanical Garden, and did graduate work at the University of Wisconsin. He has taught botany at the University of Illinois since 1932, except for the period 1942–1945, when he served the U.S. government as a rubber specialist in South America. Professor Fuller has made research contributions to the study of effects of radiations on plants, plant ecology of South America, photoperiodism, and natural rubber production. His published books include *The Plant World,* a text for one-semester botany courses, and *The Science of Botany* (with a co-author) for one-year courses in botany.

Donald D. Ritchie is a graduate of Furman University and received the M.A. and Ph.D. degrees from the University of North Carolina. He has taught at Furman University and West Virginia University. Since 1948, he has been teaching at Barnard College of Columbia University, and is now Professor of Biology and Chairman of the Biology Department. In addition to his academic work, Professor Ritchie has been engaged in research programs at the U.S. Naval Research Laboratory and has directed its Tropical Exposure Station in Panama. A Fulbright lecturer in marine biology (1967–1968) at the University College in Galway, Ireland, his research interest is in the growth and distribution of fungi, especially saltwater fungi.

General Botany

Fifth Edition

Harry J. Fuller

Donald D. Ritchie

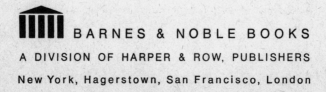

BARNES & NOBLE BOOKS

A DIVISION OF HARPER & ROW, PUBLISHERS

New York, Hagerstown, San Francisco, London

Preface to the Fifth Edition

The 1967 edition of *General Botany* has been prepared with the same aim as the older editions, but with new material. The book's purpose is timeless: to give students an overview of plant study, emphasizing its critical ideas and explaining, as directly as possible, its main topics, especially those which might seem confusing. The sections which have been rewritten in this edition are those concerned with ultrastructure, photosynthesis, respiration, growth and its regulation, DNA and its relation to genetics, and photoperiodism.

The information available since the last edition has been gained largely through the application of new techniques, especially those of chemistry and physics: chromatography, spectrophotometry, X-ray analysis, ultracentrifugation, radioactive tracers, and electron microscopy. New information on the structure and activities of plants (and indeed of all living things) includes the nature of chromosomes and especially their nucleic acids, the manner of formation of proteins, energy-transferring mechanisms in chloroplasts and mitochondria, and timing mechanisms, particularly with respect to day length.

In addition to including recent discoveries, this new edition calls attention to the many gaps in man's knowledge of plants. Students should watch for hints to tomorrow's investigators. The hints are scattered throughout the text, and say essentially: "No one knows . . . ."

I owe a debt of gratitude to Alice Ritchie for preparing the index, and to Mrs. Kayla Cohen of the editorial staff of Barnes and Noble, who has done all the things thoughtful editors should do.

——D. D. R.

Preface to the Fourth Edition

This book, like others in the College Outline Series, is a condensed study-guide, intended for use with standard textbooks in the field. The sequence of topics is that usually followed in college courses of elementary botany. Such arrangement facilitates the use of this outline as a supplement to any course regardless of the particular text being used.

This Outline is in no sense a textbook in itself. In an introduction to another book in this same Series, Professor C. C. Crawford of the University of Southern California described the proper use of Outlines as follows: "As a type of educational literature, the syllabus, or outline, has a distinct service and value, if properly used. It gives order, organization, and perspective to a field of study. It gives definiteness, objectivity, and tangible qualities to the subject or course. It provides the student with something to which he can cling. It enables the teacher and the student to define the limits of the course. The teacher can know when the material has been covered completely; the student has a checking guide by which to judge the adequacy of his study and preparation."

The author wishes to thank Mary L. Fuller and Wilson N. Stewart, who prepared most of the diagrams and drawings. Thanks are due Dr. Gordon Alexander for permission to use some of the illustrations in his *General Biology*, published by Barnes and Noble, Inc. The author acknowledges with thanks the criticisms by Dr. Oswald Tippo of certain chapters in the book.

——H. J. F.

Urbana, Illinois

Contents

x   Contents

General Botany

Chapter 1

The Study of Botany

Botany is the branch of biology which deals with the structure, physiology, reproduction, evolution, diseases, economic uses, and other features of plants. The word *botany* can be traced to ancient Greek words meaning "graze," "plants," and "cattle."

Any definition of plants must be modified by the phrase "with some exceptions," but broadly speaking, plants are organisms which produce spores and which have cellulose, a complex carbohydrate, as a component of their cellular coverings. The essential feature of green plants is their ability to synthesize their own metabolic compounds from simple substances—salts, carbon dioxide, and water.

HISTORY OF BOTANY

Ancient Period. The Greeks studied plants primarily for practical reasons, but did some speculating on the nature of plants. Aristotle fancifully compared plant nutrition with animal nutrition, but his pupil, Theophrastus (died in 287 B.C.), wrote an *Enquiry into Plants* which is still interesting. The Romans added little to botanical knowledge.

Medieval and Renaissance Periods. During the Middle Ages, most botanical study was carried on in monasteries and in gardens which were kept by wealthy men and later by universities. The study was often a mixture of observations upon plant structure and behavior, with accounts of the superstitions about and the mythology of plants. Emphasis was placed upon the study of edible and medicinal plants and frequently upon descriptions of the forms of plants. The chief botanical books, called herbals, usually contained drawings or wood-cut illustrations of the

1

plants described. During the sixteenth century, herbals contained accounts of hundreds of plants from Asia and the Americas. Some of these books, such as those by Bock, Brunfels, Gerarde, and Fuchs, were carefully and beautifully illustrated by skilled craftsmen. Simple attempts at the classification of plants were made from time to time, but no effective system was devised.

Modern Period.   The modern scientific study of the facts of plant life, divorced from superstition and mythology, began in the late seventeenth and early eighteenth centuries. Outstanding among the botanists of the early modern period was the Swedish naturalist Carolus Linnaeus (1707–1778), who established some of the principles of plant classification and named many species of plants.

As in any beginning science, the early work was observational and descriptive. Studies of classification and of gross structure were the earliest branches of botany to develop, for they required no specialized tools or techniques from other sciences. The study of the minute anatomy and the functional phases of plant life developed later, after the basic principles of chemistry and physics had been formulated and magnifying lenses had improved. Most of our knowledge of plants has come into being in the past hundred and fifty years, the bulk coming in the past century. Many major advances, especially in cellular physiology, have been made in the past two decades, and the rate of progress is constantly increasing.

During the early part of the modern period, the study of crops and gardening, and many other practical phases of plant study were considered to be fields of botany. These fields have grown so enormously, however, that they are considered separate sciences closely related to the study of botany, as we now use the word. Among these daughter-sciences of botany are *agronomy*, the science of field-crop production; *horticulture*, the science of greenhouse, garden, and orchard plants; *bacteriology*, the study of bacteria; and *forestry*, the science of the forest.

The term *botany* in its modern sense is regarded as the study of plants for the direct interest which they hold for the human mind, without any compulsion to consider practical aspects of plant life. The science of botany consists of several fairly distinct, though closely related branches:

1. *Plant Morphology,* the study of plant structure.

2. *Plant Anatomy,* a phase of morphology dealing with the minute internal structure of plants, with special reference to their tissues.

3. *Plant Taxonomy,* the study of plant classification, and the principles of plant classification and identification.

4. *Plant Pathology,* the study of the causes, control, and other features of plant diseases.

5. *Plant Physiology,* the study of the chemical and physical processes and behavior of plants.

6. *Plant Ecology,* the study of plants in relation to their environment.

7. *Plant Geography,* dealing with the distribution of plants on the earth's surface, and overlapping both ecology and taxonomy.

8. *Plant Genetics,* the study of the inheritance and breeding of plants.

9. *Plant Cytology,* the study of the structure and physiology of individual cells, especially in relation to genetics, biochemistry, and taxonomy.

The question of how plants have evolved is of primary importance in all these branches of the general science of botany.

REASONS FOR STUDYING BOTANY

An Asset in a Liberal Education. Since plants constitute one of the most conspicuous features of human surroundings, a knowledge of the principles which govern plant life is an important part of a liberal education. Plants have been part of all major human activities, and are involved in studies of anthropology, economics, esthetics, literature, religion, and politics.

An Appreciation of the Place of Human Beings in Nature. An awareness of man's dependence upon plants for food, textiles, rubber, dyes, lumber, medicines, oxygen, and many other products increases man's appreciation of the activities of plants and of his place in nature. Plants are by far the most numerous living things on earth, and man, in spite of his impact upon some parts of the earth's surface, is a newcomer who is in a relatively minor position.

An Important Tool in Many Professions.   In many practical fields, such as forestry, pharmaceutics, agronomy, horticulture, plant breeding, soil conservation, and bacteriology, a knowledge of the fundamental features of plant behavior is essential or exceedingly helpful.

Chapter 2

The Living State:
Plants and Animals

"Life" has never been defined in a way satisfactory to all those concerned with its definition. An admittedly imperfect definition, used commonly by biologists, is that life is the sum total of those phenomena exhibited by organisms. Organisms possess a combination of the following abilities which distinguish them from non-living entities:

1. *The Power of Assimilation,* or the ability to convert non-living materials into living substance (*protoplasm*) with its characteristic, specialized molecular arrangements.

2. *Irritability,* or the ability to react to the environment.

3. *The Power of Reproduction of Similar Offspring.*

4. *The Power of Reorganizing Organic Molecules,* known as foods, with the release of energy which is used in the performance of all physiological processes.

5. *The Power to Develop and Maintain a Specific, Complex Organization* of individuals, organs, tissues, cells, and subcellular parts.

THE EXPLANATION OF LIFE

Two schools of thought, the mechanistic school and the vitalistic school, attempt to explain the nature of life.

The Mechanistic Philosophy.   According to this view, all activities of organisms can be explained in terms of chemical and physical reactions. The practice of most biological research is based upon the mechanistic explanation of life and thus makes use of the tools of chemistry and physics. Though their investigations are based on mechanistic assumptions, many biologists,

5

recognizing that the fundamental nature of life has never been explained, regard vitalism with respect.

The Vitalistic Philosophy. This doctrine holds that not all activities of organisms are explainable in terms of chemistry and physics and the mathematics of chance, but that organisms also depend upon a mysterious force which cannot be measured or analyzed by the methods of science. The mechanists' objections to vitalism are that its assumptions cannot be subjected to rigorous verification, and that strict adherence to vitalistic ideas fails to stimulate experimental advancement of biological knowledge.

THE ORIGIN OF LIFE

Several theories have attempted to explain the origin of life on the earth's surface:

The Theory of Divine Creation. This view assumes that life began as a special act of God. The theory is based on faith.

The Inter-planetary Theory. The basis of this assumption is that the first life to appear on the surface of the earth reached the earth from some other planet, or from somewhere in space. This theory merely pushes the problem back to an earlier time and is not regarded seriously by most biologists.

The Theory of Spontaneous Generation. This hypothesis holds that life arose, and possibly still arises, directly from non-living matter. Pasteur and others seem to have disproved this theory, yet life obviously did begin at some time on earth. *In vitro* experiments have demonstrated that complex organic compounds can be synthesized in the laboratory under conditions which simulate those that probably existed before the development of life.

The belief that life can come from non-life is called *abiogenesis*, while the opposing belief that life can come only from life is called *biogenesis*.

PLANTS AND ANIMALS

Man has separated organisms into two groups, plants and animals. There are no differences which distinguish all plants from all animals. The fact that no single difference separates all plants from all animals is important evidence for the idea that plants and animals are related and have a common ancestry. In some primitive organisms, plant and animal characteristics are so mixed that it is impossible to classify them as either plants or animals. The intergrading of diverse groups, with some individuals sharing features of more than one category, confuses some people. A green organism which can photosynthesize food and at the same time move about and ingest food is neither an obvious plant nor an obvious animal. Classifying a higher plant (such as a flower), or a higher animal (such as a mammal) is not so confusing. These examples merely indicate that all man-made categories can have unclassifiable intermediates; yet most categories are reasonably useful, and most members of such categories can be recognized. The differences that distinguish most plants from most animals are:

1. Most plants are able to manufacture their own food from raw materials from the air and soil, while animals lack this ability and depend upon plants for their food.

2. Most plants have green pigments, *chlorophylls*, which are lacking in most animals.

3. Most plants contain *cellulose* in their structural framework, a substance lacking in all but a few species of animals.

4. Most plants are stationary, whereas most animals are capable of locomotion.

5. Most plants have an unlimited scheme of growth, as contrasted with the limited scheme of most animals. In unlimited growth, *meristems* (growth zones) persist during the life of the plant and add to its size.

6. Most plants produce *spores*, non-sexual reproductive cells, which are generally lacking in animals.

Chapter 3

The Gross Structure of Seed Plants

In spite of superficial differences, most vascular plants (plants with woody tissues) have one structural and developmental plan on which countless variations are based, whether the plant be a stubby, bristly cactus, a symmetrical Christmas tree, a slender vine, or a cabbage. (See Figs. 3/1 and 3/2.) The stem may elongate enormously and increase little in girth, or it may scarcely elongate at all and spread laterally. The leaves may enclose the entire plant, be restricted to the base of the stem, or dry up immediately after being formed and never function. Nevertheless, the stem is essentially a cone-shaped object, whether it be a short, fat one or a long, thin one, and whether it be geometrically regular or contorted.

A plant's final form, and indeed the final form of all multicellular organisms, is determined by the direction of its cell divisions, by the number of cell divisions in a given direction, and by the subsequent enlargement of newly produced cells.

GROWTH REGIONS

In seed plants, the actively growing regions are restricted to certain perpetually growing places: the tips of stems (the *apices*, singular, *apex*); *lateral* or *axillary buds*, which grow at *nodes* from which leaves are produced; *root apices* at the tips of roots; and a layer, the *cambium*, which lies immediately outside the woody tissue. Growth of an apex results in increased length; cambial growth results in increased girth; lateral buds give rise to branches. All growing regions are said to be *meristematic* (Fig. 3/1).

One large group of plants, the grasses, have a meristematic

region above the node. A stem with such a region would be too weak to remain erect if the leaves below each node did not encase the soft portion with a firm, tubular sheath (Fig. 3/2). Plants with nodal meristems lack cambium.

MAIN ORGANS OF PLANTS

The bodies of seed plants are made up of four parts: *roots, stems, leaves,* and *cones* or *flowers.* Roots, stems, and leaves are

Fig. 3/1. Diagrammatic representation of a plant with a cambium layer and other meristematic regions where cell divisions occur.

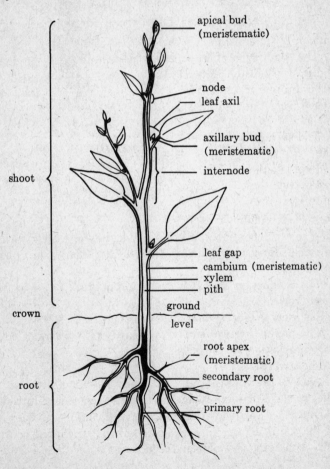

vegetative parts concerned chiefly with non-reproductive activities. Cones and flowers are the *reproductive* parts; they serve primarily for production of the *seeds,* the characteristic reproductive structures of seed plants.

Roots. Usually, but not necessarily, roots grow beneath the surface of the soil. The functions of roots are:

1. The absorption of water and dissolved materials (chiefly mineral salts) from the soil.

2. The anchorage of the plant.

3. The conduction of water and dissolved substances from the root up into the stem and of foods from the stem down into the root.

4. The storage of food and water.

5. Reproduction.

6. Photosynthesis in a few species.

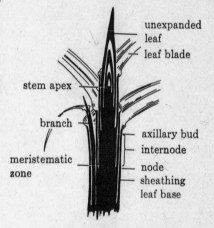

Fig. 3/2. Diagrammatic representation of a grass plant lacking cambium, but having nodal meristems protected by sheathing leaf bases.

Stems. Normally, but not necessarily, stems grow above the ground as aerial continuations of the root system. A stem with its leaves and branches is called a *shoot.* The primary functions of stems include:

1. The conduction of materials from roots to leaves and from leaves to roots.

2. The storage of foods and other substances.

3. The production and support of leaves and flowers or cones.

4. Reproduction.

5. Photosynthesis.

Leaves. Leaves are usually bifacially flattened, occasionally needle- or scale-like appendages of stems. Their chief function is the manufacture of foods from water and mineral salts absorbed from the soil and from carbon dioxide of the air. A leaf consists usually of a stalk (*petiole*) and a flattened *blade.*

Cones and Flowers.   These structures are highly modified reproductive shoots concerned with the formation of seeds. They bear specialized organs such as cone scales, stamens, etc., which perform the reproductive functions.

Chapter 4

The Microscopic Structure of Plants: Cells and Tissues

In their work on plants, botanists were limited to studying gross morphology until the development of magnifying lenses. Robert Hooke, an English scientist, in 1665 discovered tiny compartments in cork and called them *cells*. In 1675 an Italian physician, Malpighi, published an account of the microscopic, internal structure of plants. Other botanists and zoologists continued the work of examining the cellular structure of plants and animals. Several of these investigators found in the boxlike cells of plant and animal bodies a rather viscous liquid which was termed *protoplasm* by von Mohl in 1846. This substance was recognized as the actual living matter of cells. In 1759, Wolff set forth the *cell theory*, which states that the bodies of all plants and animals are composed of structural units called cells, that plants and animals develop as a result of the formation of new cells, and that the activities of an organism are a summation of the activities of its component cells. The botanist Schleiden and the zoologist Schwann, writing in 1838, were for many years erroneously credited with developing the cell theory. This theory, which was well developed by the early 1800's, summarized the growing biological conviction that the boxlike cells with their living bits of protoplasm are the units of which the bodies of all plants and animals are made. This theory has dominated most fields of biological thought since that time.

About 1885–1895 another theory of the organization of living beings—the *organismal theory*—was proposed. This maintains that an organism is to be regarded not simply as an aggregation of individual cells, but as a functional unit, subdivided for physiological and structural convenience into small masses of protoplasm which, with their surrounding walls or membranes, are called cells. The organismal theory, which has become increasingly accepted, emphasizes the idea of a coordinated whole made up of structural units which cooperate in their functions and their development.

An organism which consists of a single cell is termed a *unicellular organism*. Often a number of similar unicellular organisms become grouped together without any division of labor; these aggregations are referred to as *colonies*. A *multicellular organism* is composed of many cells which vary in structure and function.

Cytology, the study of cell structure and physiology, has developed during the past 100 years. In 1831, Brown discovered *nuclei* (singular, *nucleus*) in epidermal cells of orchids. *Chromosomes* in *Tradescantia* pollen were illustrated by Hofmeister in 1848, although he did not realize what he was observing. In the 1860's, *mitosis* (nuclear division) was elucidated by a number of investigators, especially von Beneden, Flemming, Boveri, and Strasburger. Soon thereafter, chromosome reduction (*meiosis*) was discovered, and by the end of the nineteenth century, most cellular particles were known.

The discovery of the relation of chromosomes to inheritance gave a new impetus to the study of cells during the early twentieth century. McClung, Wilson, and Morgan were some of the chief investigators in the new field of *cyto-genetics*. Meanwhile, chemical knowledge of cellular activities was also increasing, and the basic information of biochemistry was established by about 1940. Since then, enormous advances in cellular physiology have been made with the aid of new techniques, especially those employing radioactive isotope tracers, chromatographic separation of minute quantities of compounds, high-speed centrifugation, electrophoresis, X-ray analysis and electronic computers, and the electron microscope. Light microscopes cannot resolve particles smaller than about 0.25 microns (1/100,000 of an inch), but with metallic shadowing and ultra-thin sections, the electron microscope can resolve particles so small that not microns but Ångstrom units (1/10,000 microns) are used as units of measure for membranes, small subcellular particles, and macromolecules.

Since 1950, major advances have been made in research on the nucleic acids (Watson and Crick and others), the synthesis of proteins, the "genetic code" (Ochoa and others), cytoplasmic control by nuclei, photosynthesis (Calvin, Arnon, and others), and the ultrastructure of cells.

At the present time, knowledge of cellular morphology seems to be almost complete, but in spite of recent advances in research on cellular function, some of them dramatically fast, man still faces enormous gaps in his understanding of total cell operation.

CELL SIZE

Most plant cells are so small that they can be seen clearly only with the aid of a microscope. The diameters of most plant cells lie between 0.1 and 0.01 millimeters (100 and 10 microns = 1/250 and 1/2500 inches). Some cells are larger or smaller than these dimensions. Bacterial cells may be 0.5 microns (1/50,000 inches) or less in diameter, and some fiber cells in palms attain a length of several feet, but such dimensions are extremes. The small size of plant cells is illustrated by the fact that an apple leaf contains about fifty million cells.

CELL STRUCTURE

Plant cells usually have (1) a *cell wall*, (2) *protoplasm*, the living part of the cell, enclosed by the cell wall, and (3) *inclusions*, non-living structures, such as vacuoles, present in the protoplasm (Figs. 4/1 and 4/2).

The Cell Wall. The cell wall gives support and form to the cell and to the plant as a whole, and protects the protoplasm within. Between adjacent cells is an *intercellular layer* or *middle lamella*, composed chiefly of sticky, gelatinous carbohydrates known as *pectic substances*. These form a mucilaginous layer which aids in holding the other wall layers together. Inside the middle lamella are other wall layers, composed chiefly of *cellulose*, a carbohydrate which is strong, tough, and elastic and which is the most abundant constituent of the cell walls of most plants. It is, in fact, the most abundant organic compound in the world. In addition to cellulose and pectic substances, cell walls contain minerals, often waxy substances (e.g., *suberin* in cork-cell walls),

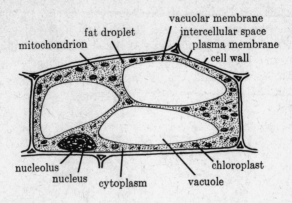

Fig. 4/1. Diagram of a mature, photosynthetic cell (magnification about 500 ×).

lignin, oils, resins, etc. Lignin is a term applied to a number of tough, complex, phenolic compounds associated with cellulose, especially in wood cells and fibers. All these substances are secreted by the living protoplasm inside the wall and are non-living. Cell walls do not dissolve in water, but are capable of absorbing and holding water in large quantities.

Cell walls are not continuous but frequently possess thin areas (*pits*) and perforations, or *plasmodesmata* (singular, *plasmodesm* or *plasmodesma*). Through the perforations, strands of protoplasm extend from one cell to another, facilitating the transfer of materials and irritable impulses. Through the pits, water and dissolved substances diffuse from one cell to another.

Protoplasm. Because "protoplasm" and "life" cannot be adequately defined, both terms are considered meaningless by some biologists. They are useful, however, even though they are vague, especially if their limitations are kept in mind. Protoplasm is the structurally complex, constantly changing aggregation of materials which fills cells, and is classically known as the "living substance." It is opalescent or transparent, and variously viscid, being sometimes almost as fluid as water and at other times as firm as rubber. It is usually grayish, but may contain red, green, yellow, blue, or black pigments. Protoplasm consists of a watery solution of salts and organic compounds, integrated with structures which can be demonstrated by means of such special techniques as the light microscope, the electron microscope, homogenation and centrifugation, etc. In most cells, the bulk of the protoplasm is the *cytoplasm*, which comprises all the metabolically active protoplasm outside the *nucleus*.

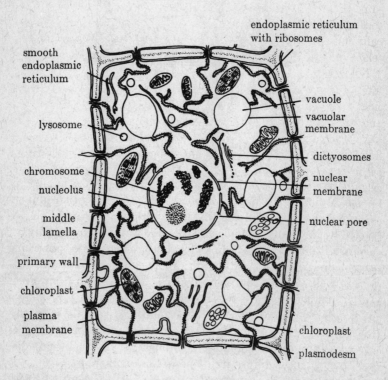

Fig. 4/2. Diagrammatic view of a thin section of a plant cell as seen in electron micrographs (magnification about 10,000 ×). Chloroplasts are shown as if they were cut through two different planes.

THE NUCLEUS. A living *nucleus* in a vegetative state (i.e., not undergoing division) can be seen under the microscope as a grayish body, spherical, elongated, or variously flattened or lobed. Most nuclei are surrounded by a porous *nuclear membrane*, and are filled with fluid *nuclear* sap in which float the *chromosomes* and one or more *nucleoli*. In some phloem cells, nuclei disintegrate before the cell dies, and some microorganisms, such as bacteria and blue-green algae, have no membrane-bound nuclei.

Chromosomes, made of nucleic acids and simple proteins, are responsible for genetic and metabolic control, and they regulate most of the syntheses and reproductive activities of cells. They are not usually visible in a living cell except when the nucleus is in process of division, at which time they may be seen as flexible rods. Chromosomes are most easily observed after being chemically hardened and stained, and were so named because of their relatively great affinity for dyes.

Nucleoli also contain nucleic acids and proteins, but function as intermediates in protein synthesis, and are chemically and functionally different from chromosomes. Interaction between nucleus and cytoplasm is essential to the normal behavior of cells.

CYTOPLASM. The cytoplasm is non-nuclear protoplasm. Its fluid sap is bounded by a sub-microscopically thin *cytoplasmic,* or *plasma,* membrane, which in higher plants lies just inside the cell wall. A labyrinthine system of complex, double membranes branches and re-branches throughout the cytoplasm, making connections with the plasma membrane, the nuclear membrane, the *vacuoles,* and probably other parts of cells. This membrane system, the *endoplasmic reticulum,* is studded with submicroscopic particles, *ribosomes,* which are made of nucleic acids and proteins, and are centers of protein synthesis. The endoplasmic reticulum extends from nucleus through cytoplasm in any one cell, and from one cell to the next through the protoplasmic connections (the plasmodesmata). A plant is therefore a unit, all of whose cells have continuous interconnections which make the plant an integrated, protoplasmic entity, not merely a collection of individual cells. Nevertheless, one cell may possess all the faculties for generating an entire plant. A carrot plant, for example, with root, shoot, flowers, and seeds, can be grown from a single phloem cell.

Vacuoles. These structures constitute the bulk of most mature cells of higher plants. Vacuoles are fluid-filled enlargements of spaces between the two layers of the endoplasmic reticulum, and thus are membrane-bound droplets. They can swell, coalesce, and come to fill as much as nine-tenths of the volume of a cell. They contain mostly water with dissolved pigments (frequently red or blue), mineral salts, organic acids, proteins, crystals, and an enormous array of various metabolic products, most of which serve functions as yet unknown. They are the main water-storage centers of cells, and their plumpness or turgor keeps cells, tissues, and organs firm.

Plastids. These structures are peculiar to plants, and contain the photosynthetic pigments. Structurally, plastids are usually somewhat flattened, ellipsoid bodies, although in algae they exist in a variety of shapes. They are membrane-bound, with a complex of submicroscopic inclusions, *grana*, which are piled up like short stacks of coins. The grana in turn have membranes, layers of protein and fatty substances, photosynthetic pigments (the chlorophylls and associated carotinoids), and the enzyme systems which make photosynthesis possible. Green plastids are *chloroplasts*. Plastids with yellow or red pigments (carotene in carrots, lycopene in tomatoes) are *chromoplasts*. Any colorless plastid can be called a *leucoplast*. Other plastids store starch (*amyloplasts*) or oils (*elioplasts*). Photosynthetic bacteria and blue-green algae lack chloroplasts, but their chlorophylls and other pigments are located in submicroscopic particles (*chromophores*).

Mitochondria. These granular or sausage-shaped inclusions are generally smaller than plastids, colorless, flexible, expandable, and motile. A double membrane covers each mitochondrion; the outer part is smooth and continuous, and the inner membrane is expanded to make wrinkles (the *mitochondrial crests*) that form incomplete partitions inside the mitochondrion. The main function of mitochondria is the formation, by means of elaborate enzyme systems, of energy-rich compounds, especially adenosine triphosphate (ATP). They also are probably concerned with the synthesis of fatty compounds, and are thought to be involved in a number of other metabolic activities, including the synthesis or accumulation of rubber particles. Mitochondria are numerous in young cells, but less common in older ones, being relatively rare in mature photosynthetic cells.

Lysosomes. These minute, membrane-bound droplets in the cytoplasm contain concentrations of various enzymes. They act as cellular scavengers, digesting cytoplasmic particles such as mitochondria which have passed usefulness. They also may bring about dissolution of entire cells, as in the liquefaction of some plant parts such as, for example, the petals of spiderwort, *Trades-cantia,* or the gills of inky-cap mushrooms, *Coprinus.*

Golgi bodies, or dictyosomes. These stacks of double-membrane-bound plates are not distinctly visible with the light microscope. Because they are concerned with cellular secretions in animals, they are thought to have a similar use in plants. At the present time, however, we understand best their activity in association with new cell walls when they are laid down as a partition between two newly divided cells.

Flagella and cilia. Both are threadlike cytoplasmic extensions which are the locomotor organelles of swimming spores or sex cells. These two structures are similar to one another except that cilia are short, proportioned about like eyelashes, and flagella are longer. They occur in many lower plants and in sperms of a few seed-plants.

Centrosomes. Common in animal cells, centrosomes are rare in plants, being found in only a few lower plants. They are short, fibrous cylinders about 0.25 microns in diameter, and are involved, in some unknown way, in nuclear division and in the formation of cilia and flagella.

Membranes. These elastic, double layers of protein and fatty materials are frequently about 75 Ångstroms thick. They form special surfaces over almost all subcellular particles as well as over entire cells. They are such essential and common features of protoplasmic entities that a cell may almost be thought of as an intricately organized collection of membranes.

TISSUES

Cells vary greatly in size, structure, and function. Their differences in structure reflect the different functions which they perform in the life of a plant. A group of structurally similar cells performing the same function is a *tissue;* a major portion of a plant body, composed of various kinds of tissues is an *organ* (e.g., a leaf or a stem). Plant tissues (Figs. 4/3 and 4/4) are broadly classified, on the basis of their structural and physiological differences, into two groups—*embryonic (meristematic) tissues,* and *permanent tissues.* The permanent tissues are further subdivided as follows:

Simple Permanent Tissues

1. Epidermis.   2. Parenchyma.   3. Sclerenchyma.
          4. Collenchyma.   5. Cork.

Complex Permanent Tissues

1. Xylem.   2. Phloem.

Fig. 4/3. Diagrams of simple permanent plant tissues (magnification about 250 ×).

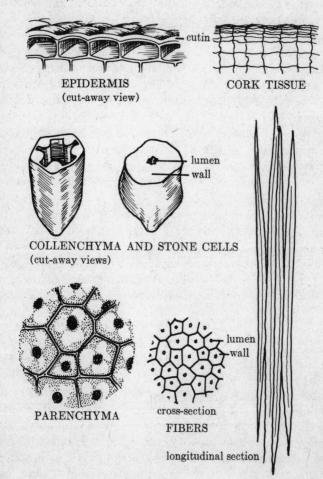

EPIDERMIS
(cut-away view)

CORK TISSUE

COLLENCHYMA AND STONE CELLS
(cut-away views)

lumen
wall

PARENCHYMA

FIBERS

cross-section

lumen
wall

longitudinal section

Meristematic Tissues.   These tissues are located at the tips of roots and stems, between the water- and food-conducting tissues of stems, and at various other places in plant bodies. Meristematic cells are small, thin-walled, frequently cubical or brick-shaped, densely packed with protoplasm, and capable of producing new cells by cell division. These new cells by processes of growth, enlargement, and differentiation become the permanent tissues of the plant. Permanent tissues do not normally become changed into other kinds of tissues as do meristematic tissues.

Permanent Tissues.

SIMPLE PERMANENT TISSUES.   Each simple tissue is composed of similar cells.

Epidermis. The epidermis covers leaves, the softer portions of stems and roots, etc. It is usually one cell thick and often has a waxy substance (*cutin*) deposited on its outer walls. It functions primarily to protect the inner tissues against excessive evaporation of water, and, in roots, to absorb materials from soil. The epidermis of leaves, and often of stems, contains pores (*stomata*), through which exchange of gases may occur. Epidermal cells are usually colorless, with the exception of *guard cells,* which contain chloroplasts. Each pair of guard cells encloses a stoma.

Parenchyma. This common tissue is composed of relatively thin-walled, large-vacuoled, cylindrical, prismatic or roundish cells with flattened faces. A mathematically ideal parenchyma cell would have a fourteen-sided body with eight six-sided and six four-sided, non-plane faces, but such a shape is rarely found in actual tissues. Parenchyma cells frequently contain chloroplasts and thus manufacture food. Non-green parenchyma cells usually store food and water.

Sclerenchyma. Sclerenchyma cells are thick-walled, strengthening cells of two kinds: (1) elongated, tapering *fibers,* and (2) short *stone cells.* Both are common in stems; stone cells occur in walnut shells, the fruit of pears, bark, etc. Full-grown sclerenchyma cells are dead.

Collenchyma. The walls of collenchyma cells have reinforcing thickenings in their corners. They are common in stems, especially herbaceous ones.

Cork. Cork cells, which have walls waterproofed by waxy *suberin,* die soon after they are formed. Occurring in bark, potato

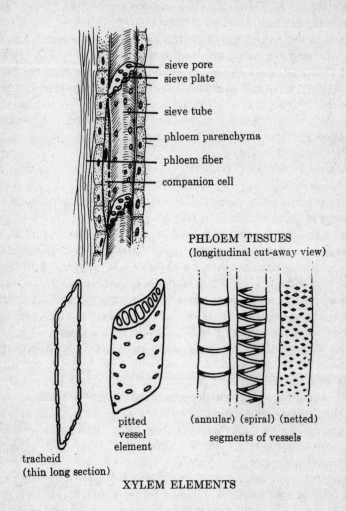

PHLOEM TISSUES
(longitudinal cut-away view)

XYLEM ELEMENTS

Fig. 4/4. Diagrams of complex permanent plant tissues (magnification about 250 ×).

skins, etc., they protect inner tissues against excessive evaporation.

COMPLEX PERMANENT TISSUES.   Each complex tissue is composed of more than one kind of cell.

Xylem. Xylem conducts water and dissolved substances chiefly upward through roots, stems, leaves, and flower stalks. The kinds of cells commonly found in xylem are *tracheids, vessels, fibers,* and *parenchyma.* Tracheids are elongated, tapering cells which are dead at maturity and which serve for strength and conduction; their walls are usually pitted. Vessels are not single cells but are long, continuous tubes formed by the dissolution of the end walls of vertically elongated cells; vessels serve chiefly for conduction. Xylem fibers are elongated, pointed, strengthening cells with much-thickened cell walls; at maturity they contain no protoplasm. Fibers differ from tracheids chiefly in their thicker walls and reduced pits. Xylem parenchyma cells are storage cells. In some plants, the xylem lacks one or more of these kinds of cells.

Phloem. Phloem conducts chiefly foods, principally downward from leaves into stems and roots. It is composed of *sieve tubes, companion cells, fibers,* and *parenchyma.* Sieve tubes are vertically elongated rows of cylindrical cells with perforated end walls and cytoplasm; these are the chief conducting cells of the phloem. Companion cells border the sieve cells and seem to aid them in conduction. Phloem fibers are thick-walled, elongated, strengthening cells. Phloem parenchyma cells are storage cells.

Chapter 5

Relationship of Plant Cells to Water

Recognition of the essentiality of one compound or other (water, protein, nucleic acids, etc.) sometimes tempts students and even trained scientists to refer to it as "most important." No compound can be most important because without the full complement of materials, the complex interplay which we call life would cease. Nevertheless, water is the one compound common to all organisms on earth, is the medium through which all biological substances move, makes up the bulk of protoplasm, and indeed is of prime importance to all living things. It consequently receives much attention from investigators.

PHYSICO-CHEMICAL NATURE OF PROTOPLASM

Protoplasm is not simply a solution in which individual molecules or ions of several kinds are uniformly distributed, but is a mixture of some substances in solution, some macromolecules, many membranes, and various aggregations of molecules in a *colloidal* state. Many colloidal materials occur in inanimate nature: fog, opals, meerschaum, mayonnaise, murky waters, smoke. In protoplasm, the suspending medium is water, and the suspended particles are usually proteins or fatty materials. The particles are large enough to settle out under the drag of gravity but are held suspended by coatings of closely packed water molecules, or by charged particles, or both. The protective layers prevent coalescence of the colloidal particles, which consequently remain floating in the suspending medium. Colloidal systems are biologically important because they can hold relatively great amounts of water and because they present a large surface area

in proportion to the volumes of the particles. Most biological reactions occur at surfaces.

Relatively few of the known chemical elements are essential to
living cells. No element occurs exclusively in protoplasm, but
many elements are passively allowed to enter and accumulate in
plants even though they serve no purpose. Some of these apparently useless elements are lead, gold, aluminum, silver, strontium,
arsenic, and selenium. The most abundant elements in protoplasm
are *carbon, oxygen, hydrogen,* and *nitrogen.* Present in smaller
quantities are *sulfur, phosphorus, iron, calcium, potassium,* and
magnesium. Some elements are required in such small quantities
that they are called *trace elements* or *micromineral elements:*
boron, cobalt, copper, manganese, and zinc. Chlorine and fluorine
are probably also essential, but iodine, commonly essential to
animals, has not been shown to be necessary to plants. Some
plants, especially marine microorganisms, have a sodium requirement.

Elements do not usually occur in protoplasm in their elemental
molecular form, but as ions or as parts of compounds.

Water is the most abundant compound in active cells, frequently
constituting more than 90 per cent of protoplasm. *Proteins* form
a large and important part of living material, but they would be
no more than inert substances without the array of *sugars, fats,
pigments,* and *organic acids* with which they must be associated.

The individual chemical compounds of protoplasm cannot be
designated as living or non-living. It is the complex system of
these chemical compounds which is alive.

Plant cells absorb large quantities of water through the mechanisms of *imbibition, osmosis,* and by *physiologically active
transport.* Water is important in plants because:

1. It is a constituent of living protoplasm.

2. It is a raw material used in food manufacture.

3. It is the medium of absorption and of transportation of
solid materials in plants.

4. It is the medium in which most of the chemical reactions in
protoplasm take place.

5. It provides the pressure which is necessary for the maintenance of form, for support, and for growth.

6. It helps regulate temperature.

Imbibition. This process entails the absorption of water by dry or partly dry colloidal materials, as, for example, the absorption of water by cotton, by a sponge, by a blotter, etc. Cell walls and protoplasm absorb water by imbibition. Cell walls of most plant cells are *freely permeable*, that is, they permit the passage of water and dissolved materials. Imbibition of water by biological materials causes them to become softer, swollen, and more elastic.

Osmosis. Osmosis is the diffusion of a liquid (water, in organisms) through a *differentially permeable membrane* (a membrane which allows certain substances to pass through it, but which restricts or prevents the passage of other substances). Osmosis depends upon (1) diffusion, (2) differences in concentration of the diffusing substance, (3) a differentially permeable membrane, and (4) the nature and concentration of dissolved materials not able to pass through the membrane.

DIFFUSION. Diffusion is the tendency of the molecules of a substance to move from a region of greater abundance (concentration) to a region of lesser abundance. Gases diffuse (e.g., ether from an open bottle), liquids diffuse (e.g., alcohol diffusing in water), solids diffuse into liquids (e.g., sugar diffusing into water).

DIFFERENCES IN CONCENTRATION OF THE DIFFUSING SUBSTANCE. The *direction* of diffusion depends upon the concentrations of the diffusing substance in different places; a substance normally diffuses from the place of its greater concentration to that of its lesser concentration. The chief tendency in diffusion is toward equalization of concentrations. The *rate* of diffusion depends on difference in concentrations (the *diffusion gradient*), temperature, and other factors, such as the size of the diffusing particles.

A DIFFERENTIALLY PERMEABLE MEMBRANE. Parchment, an example of this type of membrane, permits water to pass but impedes the passage of sugar molecules. If the large end of a funnel is covered with a parchment membrane, and the funnel is filled with sugar solution and placed with its large end down, just beneath the surface of water in a vessel, osmosis occurs. Relatively more water is in the vessel outside the funnel than in the solution in the funnel; consequently, water moves into the funnel from the vessel through the membrane in accordance with the diffusion tendency to equalize concentrations of water on both sides of the membrane. The only manner in which this tendency can be satisfied is by the passage of water into the funnel; equalization of concentration of the two solutions cannot be achieved by the diffusion of sugar out of the funnel into the vessel because the membrane is impermeable to sugar. In this demonstration, water accumulates in the funnel and if the small, open neck of the funnel is stoppered, the water begins to exert a pressure, called *hydrostatic pressure*. Hydrostatic pressure in a cell is called *turgor pressure*.

EFFECT OF SUBSTANCES UNABLE TO PERMEATE THE MEMBRANE. The greater the concentration of dissolved particles to which a membrane is impermeable, the smaller is the relative concentration of water in which these particles are dissolved. In the demonstration above, the greater the number of sugar particles in the water inside the funnel, the smaller is the relative concentration of water molecules in the funnel, and the greater the difference between the concentrations of water inside the funnel and in the vessel outside the funnel. Thus, the kind and number of dissolved particles to which a membrane is impermeable influence, on both sides of the membrane, the relative concentrations of the liquid in which they are dissolved, and hence influence osmosis.

Active Water Absorption. Imbibition and osmosis alone cannot explain all water absorption by plant cells. *Active water absorption* appears to be related to energy released in respiration, and bulk flow of water into cells is more than a simple case of molecular diffusion.

ABSORPTION OF WATER BY LIVING CELLS

A plant cell, such as a root epidermal cell, is somewhat similar to the funnel described above. The vacuole with its cell sap is a water solution of sugars, salts, and other substances, comparable to the sugar solution in the funnel. Between the cell wall and the vacuole are the cytoplasmic and vacuolar membranes, comparable to the parchment membrane. Outside the root cell is the soil solution (water with small concentrations of salts and other dissolved materials), comparable to the pure water in the vessel. The cytoplasmic membranes are differentially permeable; they allow water to pass through freely but they restrict the passage of sugars and certain other dissolved substances in the cell sap. Because of the large numbers of dissolved particles of such substances in the cell sap, relatively less water is in the cell sap than in the soil solution. As a result, water moves into the protoplasm from the soil solution.

Water moves from cell to cell in large part as a result of differences in water concentrations (or stated conversely, differences in concentrations of dissolved substances) in different cells. The term *osmotic concentration* refers to the relative number of dissolved particles to which the cell membranes are impermeable or nearly so. The direction of water movement is from cells of low osmotic concentration to cells of higher osmotic concentration. *Osmotic pressure* or *osmotic value* is the maximum pressure which could develop in a solution separated from pure water by a rigid membrane permeable only to water. The *turgor pressure* which develops in a cell as a result of water entry is less than the theoretical osmotic pressure which could develop in that cell because cell membranes are not rigid and are not ideally differentially permeable. Furthermore, phenomena other than osmosis affect water movement.

PLASMOLYSIS

Water continues to enter a plant cell so long as the concentration of water (*not* dissolved solids) in a solution outside the cell exceeds that inside the cell, but after a time an equilibrium will be reached and the rate of movement of water outward will equal the inward rate. If, however, the concentration of water in a solution outside a cell becomes less than that inside, water diffuses out of the cell in accordance with the laws of diffusion. This outward diffusion of water causes a shrinkage of the protoplasm away from the cell wall, resulting in a condition known as *plasmolysis*. Plasmolysis occurs when salt is placed about the roots of a plant or when a piece of plant tissue is placed in a salt solution. A disturbance which affects the selective permeability of cell membranes may allow dissolved materials to diffuse out of the cell, with resultant collapse of the *protoplast* (the protoplasm of a single cell).

The presence of large amounts of salt results in a decrease in the relative concentration of water outside the cells, as a result of which outward diffusion of water and plasmolysis occur. Plasmolysis is involved in the preserving of meats and jellies by adding salt and sugar, respectively, to them. The spoilage of such foods is caused by living bacteria and molds present in the air. Falling upon salted meat or upon sugared jellies and preserves, these bacteria and molds are plasmolyzed and thus spoilage is averted. The addition of excessive amounts of fertilizer salts to soils often causes plasmolysis. If plasmolysis continues too long, the death of the plasmolyzed cell results, especially if, in addition to the plasmolytic effect, the salts in the fertilizer create an *ion imbalance* with too much of one kind of ion in proportion to others. The resulting *salt toxicity* damages the cell membranes, and this damage may be irreversible.

THE ABSORPTION OF SOLUTES

The absorption of dissolved materials (*solutes*) by living cells is a complex phenomenon not yet completely understood. Diffusion is one of the controlling factors in this process. In *passive absorption,* solutes follow the fundamental law of diffusion— that is, they move from regions where they are more abundant (e.g., soil) to regions where they are less abundant (e.g., root cells). Often, however (in *active absorption*), the movement of solutes into plant cells is contrary to this law of diffusion. The phenomenon of active absorption, which proceeds contrary to the laws of simple diffusion, has not been completely explained; certain colloidal phenomena are involved and metabolic activities involving the expenditure of energy are important in bringing about active absorption. Generally, only substances dissolved in water can enter or leave protoplasm, but some ionic exchanges from soil particle to protoplast may occur directly. In most cases, the absorption of a single kind of substance is independent of the absorption of other solutes; each solute behaves independently, diffusing as described above from high to low concentration, or frequently, in active absorption, from low to high concentration of the solute.

The principal solutes that plants absorb from the soil are mineral salts: sulfates, nitrates, phosphates, etc. The various soil solutes that roots take up from the soil are absorbed chiefly in *ionic* form rather than in *molecular* form. Ions usually do not accumulate in root cells but move to other cells where they are used in food manufacture and other processes. Thus, the concentrations of these ions is lower in roots than in the soil, and the ions may pass into root cells in accordance with the fundamental laws of diffusion.

Chapter 6

Roots and their Relation to Soils

Plants are in close contact with the soil through their root systems. Most roots grow beneath the surface of the soil. The functions of roots are:

1. Anchorage of the plant in the soil.
2. Absorption of water and dissolved minerals from the soil.
3. Conduction of water and minerals upward into the stem.
4. Conduction of foods manufactured in leaves downward to the growth and storage regions of roots.
5. Food storage.
6. Reproduction.
7. Photosynthesis (in some aerial roots).

SOILS

Soils vary in their origin, chemical nature, physical properties, depth, acidity, etc. Most soils consist of (1) air, (2) water, (3) rock particles, (4) mineral salts and other soluble inorganic chemical compounds, (5) organic matter, and (6) living organisms.

Air Content.　Because oxygen is necessary for respiration, the air within soils is important to plants. In clay soils there is more air than in loam and sandy soils which consist of larger particles and have less air space. If spaces among soil particles become filled with water (waterlogged), air is forced out, creating a condition which, if it persists, may cause injury to plants. Plowing separates soil particles and results in better aeration. The respiratory activity of roots, bacteria, fungi, and animals living beneath the soil surface decreases the oxygen content of soil and increases the carbon dioxide content.

Water Content.   The amount of water in soils is determined by rainfall, drainage, the water-holding power of soil particles, the nature of the subsoil, etc. Water which flows away from the surface of soil is *runoff water*. *Gravitational water* is that which percolates down through soil to the standing water (*water table*) below the soil surface. *Capillary water* is held loosely by soil particles. It is chiefly capillary water that is available to the roots of plants. *Hygroscopic water* is that which is held tenaciously by soil particles after capillary water has been removed. Hygroscopic water cannot be removed from soil particles by roots and thus is of little importance to plants. Clay soils hold water tenaciously against both plants and evaporation.

As certain parts of the soil dry out, some water may move by capillary movement into the dry portions from moister parts of the soil, but water moving thus is relatively unimportant to plants. When water enters a root, the soil immediately surrounding the root (the *rhizosphere*) becomes drier, and since water moves relatively little in soil, the root will die unless (1) more water is added from above by rain or irrigation, or (2) the root grows into a new region which has not been depleted of its water. A method of reducing evaporation of water from soil is *mulching* —the spreading on the soil of a layer of material such as straw, dead leaves, paper, etc.—which screens soil from wind and sun and thus reduces evaporation. Another advantage of mulching (and plowing) is the reduction of weed growth.

Rock Particles.   These particles vary in size from submicroscopic, colloidal particles of clay to large sand and gravel particles. Soil particles originate from the decomposition of rock by the action of water, freezing and thawing, winds, glaciers, the action of organic acids from roots, etc.

Mineral Salts.   Nitrates, sulfates, phosphates, magnesium, potassium, calcium, etc. are raw materials used by plants in the manufacture of foods and other organic materials. These substances develop partly from the disintegration of rock particles, and become available repeatedly as plant and animal bodies die and decay. In cultivated crops, chemical fertilizers supply many of the mineral nutrients in soils. Except for carbon dioxide and some artificial addition of salts to leaves, all the materials of a plant body enter the plant through roots.

Organic Matter. The waste products and decomposing remains of plants and animals provide the soil with organic material. Much of this material is derived from the dead roots, leaves, and stems of plants. The term *humus* is applied to the partly decayed, dark-colored organic residue of soil. Most humus is a complex mixture of proteins and lignins from incompletely decomposed plants. Organic matter is important in that it promotes retention of water, prevents caking, and increases porosity and aeration. It also restores (as a result of its disintegration by bacteria, molds, and other soil organisms) to the soil and air the inorganic mineral salts and gases absorbed and used by green plants in food manufacture. Organic matter is thus important in soil fertility. The addition of manure and the plowing under of certain plants, such as alfalfa, increase the organic matter of soils. Organic matter promotes the growth of soil organisms.

Organisms. Many kinds of organisms live in the soil and may be considered a part of it. These include astronomical numbers of bacteria, fungi, algae, roots of higher plants, protozoa, worms of many kinds, insects, and various larger animals. Their waste products and dead bodies enrich the organic matter of soils, they aid in the decomposition of organic matter and rock particles, and they influence the physical properties and aeration of soils by stirring up the soil particles. Organisms also affect, and are affected by, the acidity or alkalinity of the soil solution. Acidity is measured in pH units on a scale from pH 1 (very acid) to pH 14 (very alkaline), each unit representing ten times the acidity or alkalinity of the next; pH 7 represents neutrality. On such a scale most soils have a pH of about 4.5 to 6.5, although extremes beyond this range do occur. Extreme acidity in a soil, as in bogs, is caused by the action of organisms; but extreme alkalinity, as in desert salt flats, is caused by geological and weather conditions.

THE INFLUENCE OF PLANTS ON SOILS

Plants influence soils in three general ways:

1. They absorb materials (chiefly water and mineral salts) from soils. If the fertility of soils is to be maintained, these mineral salts must be restored by the addition of fertilizers, for the harvesting of fruits, leaves, roots, and other products of plants

depletes the mineral content of soils. Soils may be fertilized by the addition of mineral salts, manure, dead plant bodies such as decayed leaves, straw, etc. Crop rotation (the planting of different kinds of crops in different years) aids in preventing too rapid loss of fertility, for different crops absorb mineral salts in different quantities and thus do not exhaust the soil of any one or several minerals as rapidly as the repeated growth of a single kind of crop.

2. Plants add materials to soils. Dead and decayed plants add to the organic matter of soils, and the roots of living plants also excrete certain materials into the soil. The chief organic substance is carbon dioxide, which with water forms carbonic acid, a substance which attacks and disintegrates certain types of rock. The roots of some plants excrete waste organic products, some of which may be toxic to the plants that produce them or to other plants. Root exudates of some plants, e.g., walnuts, inhibit the growth of other species; or they may produce secondary effects by altering the microflora or the pH of the soil, thereby changing the solubility of some ions. Conversely, some root exudates, as amino acids from corn that affect the germination of witch-weed seeds, are helpful or even necessary to the life of other plants.

3. The many-branched root systems of plants aid in holding soil, thus reducing or preventing soil erosion. Also, plants break the force of wind and rain. Grasses and forests are effective in binding soil.

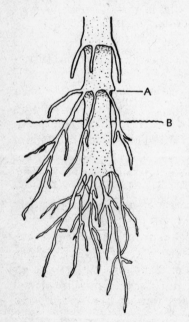

Fig. 6/1. Adventitious roots of corn. A. Node. B. Soil surface.

THE ORIGIN OF ROOTS

A seed contains a tiny plant, the *embryo*, a portion of which is

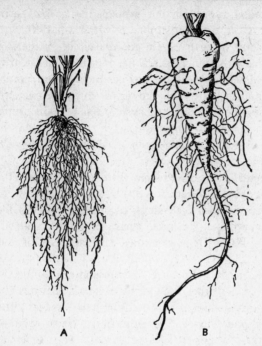

Fig. 6/2. Root systems. A. Diffuse root system (rye). B. Tap-root system (carrot).

the *hypocotyl*. This structure or a portion of it becomes the first root (*primary root*) of a plant. As it grows, the primary root produces branches, termed *secondary roots*. A primary root usually grows vertically downward, with its secondary roots in a somewhat horizontal direction. Sometimes roots arise from stems, leaves, or other parts of plants; such roots, which develop from structures other than a hypocotyl or a primary root, are called *adventitious roots* (Fig. 6/1). The prop roots of corn, the aerial roots of ivy, and the roots formed on stem and leaf cuttings are adventitious.

THE GROSS STRUCTURE OF ROOTS

The term *root system* is applied to the entire mass of underground roots produced by a plant. The extent, form, depth, and other features of root systems vary in different species of plants and under different conditions of growth.

Two common types of root systems (Fig. 6/2) are *diffuse* (*fibrous*) *root systems,* and *tap-root systems.* A diffuse root system has several or many main roots, usually slender, of about the same size, with numerous smaller root branches. Corn, wheat, and other grasses have diffuse, slender roots. Sweet potatoes and dahlias have diffuse roots, the larger of which become swollen by the storage of food. A tap-root system has a main primary root which is conspicuously longer and usually thicker than all other roots of the system. Beets, carrots, and dandelions have tap-root systems.

Root systems often exceed stem systems in their length, degree of branching, etc. Alfalfa roots are sometimes 40 feet in length, those of sugar beets 5 feet, those of black locusts 50 feet. In a full-grown rye plant, the total length of roots of all sizes may reach 380 miles, with a combined surface area of over 2500 square feet.

Roots are usually cylindrical in form and are usually colorless, or have some color other than green. Roots lack *nodes* ("joints") and *internodes* (distances between successive nodes). Roots lack buds, leaves, and flowers. The branches of roots originate from an internal tissue (pericycle) of the root; they do not arise externally (from buds) as do the branches of stems. Young roots bear root hairs, fragile projections from the epidermal cells of the roots; these hairs appear white and cottony to the naked eye. They increase greatly the absorptive surface of the root. It has been estimated that the root hairs of a rye plant have a total length of about 6600 miles. The root hairs do not grow at the very tip of a root. This region is covered by the *root cap,* a thimble-shaped mass of cells which covers the *embryonic* tissue and thus protects it from injury from soil particles.

THE MICROSCOPIC STRUCTURE OF ROOTS

A knowledge of the internal anatomy of roots is obtained from the examination of (1) a longitudinal section of a root, (2) a cross-section of a root.

Longitudinal Section of a Young Root. In such a section, four cell regions of somewhat different aspect are visible: the root cap, the meristematic region, the elongation region, and the maturation region. (Fig. 6/3.)

ROOT CAP. This thimble-shaped mass of moderate-sized cells forms the apex of the root and protects the *meristematic* (dividing) calls just above it. The slimy outer cells of the root cap are continually being broken off by their contact with rock particles of the soil. As the outermost cells of the root cap are sloughed off, new root cap cells are being formed in the inner part of the root cap by cells of the meristematic region.

MERISTEMATIC REGION. This area comprises a mass of small, nearly cubical cells with thin walls and dense protoplasm. It is the region in which new cells are formed by mitosis and in which, therefore, the first phase of growth of a root in length is brought about.

ELONGATION REGION. The mass of cells here was recently formed in the meristematic region and is undergoing enlargement, particularly in length. In this region, cell walls increase in length, new protoplasm is formed, and vacuoles increase in size. The elongation and meristematic

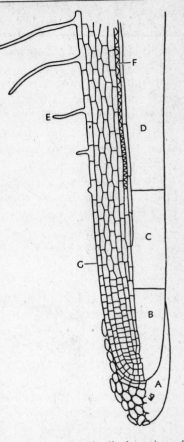

Fig. 6/3. *Longitudinal section of young root. A. Root cap. B. Meristematic region. C. Elongation region. D. Maturation region. E. Root hair. F. Stele. G. Epidermis.*

regions together usually do not exceed a few millimeters in length.

MATURATION (DIFFERENTIATION) REGION. This region is situated above the elongation region. In this region, the enlarged cells become differentiated into the mature tissues of the root—xylem, phloem, etc. All the portions of the root above the elongation region may be included in the term *maturation region*. The younger part of the region of maturation is the *root hair* zone, in which the epidermal cells develop protuberances known as *root hairs.* Root hairs are thus merely evaginations of epidermal cells; they rarely exceed a few millimeters in length or live more than a few days or weeks. New root hairs develop at the lower end of the root hair zone, usually at about the same rate as older root hairs die at the upper limit of the root hair zone. Root hairs become entwined among and closely appressed to soil particles. If a plant is pulled up by the roots, most of the root hairs are broken off by the soil particles. Most of the materials absorbed by roots are absorbed by the root hairs, which may increase the absorptive surface up to 20 times.

Cross-Section of a Root Through the Region of Maturation. In such a section the following tissues are visible: epidermis, cortex, pericycle, xylem, phloem, and parenchyma. (Fig. 6/4.)

Fig. 6/4. Cross-section of a young root (Smilax). *A. Epidermis. B. Hypodermis. C. Cortex. D. Endodermis. E. Pericycle. F. Xylem. G. Phloem. H. Parenchyma tissue.*

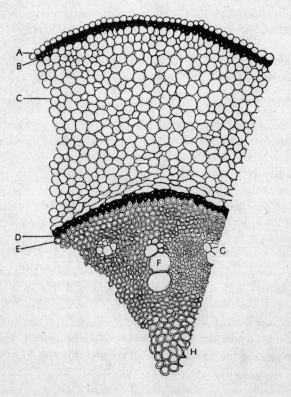

EPIDERMIS.   This surface layer of cells absorbs water and dissolved materials from the soil and offers some protection to the inner tissues of the root.

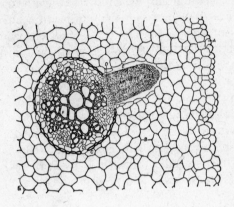

Fig. 6/5. Cross-section of young root in region of maturation, showing origin of branch root. A. Epidermis. B. Cortex. C. Endodermis. D. Pericycle. E. Branch root. F. Xylem G. Phloem. H. Parenchyma cells.

CORTEX.   This region comprises somewhat irregularly-shaped parenchyma cells with many intercellular spaces. The cortex is chiefly a water- and food-storage region. The innermost cell-layer of the cortex is the *endodermis,* which usually has its inner and side (radial) walls thickened with *suberin,* a waterproof material. The endodermis apparently functions as a water dam which prevents the outward passage of water from tissues inside the endodermis. The endodermis is believed by some to play a part in the development of root pressure (see p. 59).

PERICYCLE.   This layer of cells lies interior to the endodermis. By cell division, the pericycle gives rise to branch roots, which force their way out through the cortex and epidermis (Fig. 6/5).

XYLEM.   Composed of vessels, tracheids, etc., xylem conducts water, minerals, and often foods upward. Xylem in a root is built like a fluted column with various numbers, two to many, of flutings. If the root has four such ridges, they appear in cross-section as arms of a cross (Fig. 6/5). Branch roots originate at points in the pericycle where the radial "arms" of the xylem reach the pericycle. As branch roots mature, connection is made between the new xylem of the new root and the old xylem of the old root.

PHLOEM.   The phloem is composed mainly of sieve-tubes and companion cells in groups which alternate with the radial xylem bands. It functions chiefly for the downward conduction of food.

PARENCHYMA. Surrounding the bands of xylem cells and phloem cells are the parenchyma cells which store food and give support to these other tissues.

The term *stele* is applied to the central cylinder, which includes the pericycle, xylem, and phloem tissues.

As roots grow older, a *cambium* often develops between the xylem and phloem areas. This covering of cells over the xylem is made of meristematic cells which by periclinal cell division produce *secondary* xylem and phloem and thus increase the diameter of the root.

THE PHYSIOLOGY OF ROOTS (SUMMARY)

1. Absorption of materials occurs by osmosis, imbibition, diffusion of solutes, and active absorption.

2. These materials then pass from the root hairs through the cortex, endodermis, and pericycle into the xylem cells, which then conduct the materials upward.

3. Storage of food and water occurs chiefly in the cortex cells of roots and, to a lesser extent, in the parenchyma cells of the stele.

4. Growth in length occurs in the tips of roots, in diameter in the cambium between the xylem and phloem.

5. Reproduction is brought about in some plants by adventitious buds formed on roots.

6. Roots give off waste products into the soil.

SPECIALIZED ROOTS

The roots of some plants perform functions other than those typical of most roots or perform only a portion of these functions in specialized degree. Such roots are termed *modified* or *specialized roots*, among the common types of which are:

AERIAL ROOTS—e.g., in ivy—give support to the climbing stems, may absorb some moisture.

PROP ROOTS of corn and pandanus give support to stems.

STORAGE ROOTS of carrots, beets, etc., are enormously swollen with stored food, and enable plants to survive cold or dry seasons, or to reproduce vegetatively.

CONTRACTILE ROOTS of bulbs and other underground stems often contract in growth and pull the bulbs deeper into soil.

PHOTOSYNTHETIC ROOTS of some orchids have chloroplasts.

Chapter 7

The Gross Structure of Stems

As we mentioned earlier, the chief functions of stems are: (1) the conduction of materials, (2) the production and support of leaves and reproductive structures, and (3) the storage of food. In some plants, stems have a few other functions of a more specialized nature.

THE ORIGIN AND NATURE OF STEMS

The first stem of a seed plant develops from a part of the seed embryo known as the *epicotyl*, which is a continuation of the hypocotyl. The epicotyl is a cylindrical structure with a mass of meristematic cells and often a pair of small leaves at its apex. In some plants only the epicotyl appears above the ground; in other cases the epicotyl together with a part of the hypocotyl emerges above the ground. In the former case, the stem develops entirely from the epicotyl (e.g., in garden peas); in the latter, the stem develops chiefly from the epicotyl, with its lower part originating from the hypocotyl (e.g., in garden beans).

A stem with its leaves is termed a *shoot*. All the stems, branches, and leaves of a plant constitute its *shoot system*. Most stems grow above ground (*aerial stems*); some grow underground (*subterranean stems*). The aerial stems of most plants are *erect* (e.g., elm); in some plants they are *climbing* (e.g., morning-glories), and in others they are *prostrate* (e.g., cucumbers). Subterranean stems also show a variety of growth habits.

THE EXTERNAL STRUCTURE OF STEMS

Aerial stems are commonly classified into two types: (1) *Herbaceous stems*. (2) *Woody stems*.

These types of stems differ chiefly in the following ways:

Herbaceous Stems	Woody Stems
1. Soft and green	1. Tough and not green
2. Little growth in diameter	2. Considerable growth in diameter
3. Tissues chiefly primary	3. Tissues chiefly secondary
4. Chiefly annual	4. Chiefly perennial
5. Covered by an epidermis	5. Covered by corky bark
6. Buds mostly naked	6. Buds chiefly covered by scales

A *tree* is a woody-stemmed plant with a single main stem (trunk). A *shrub* is a woody plant with several stems of about the same size. Woody plants about twenty feet or more in height are arbitrarily called trees, even if they have multiple stems.

Buds and Leaves.   An examination of a stem in active, growing condition shows a variety of structures, the most common of which are *buds* and *leaves*. Buds are located in the axils of leaves, that is, in the upper angles between the points of juncture of leaves with stems. A bud located at the tip of a stem or twig is a *terminal* or *apical bud;* a bud located along the side of a stem is termed a *lateral,* or *axillary bud.* Buds sometimes develop in places other than the axils of leaves; such buds are *adventitious.* The point on a stem from which a leaf or bud grows is called a *node;* the length of stem between two successive nodes is an *internode.* (Fig. 3/1.)

Buds may be classified in several ways, as follows:

1. *Terminal* and *axillary buds.*
2. *Naked buds* (without scales) and *covered buds* (with scales).
3. *Active buds* and *dormant* (resting) *buds.*
4. *Flower buds* (produce flowers), *branch* or *leaf buds* (produce leaves), and *mixed buds* (produce both flowers and leaves).
5. *Alternate buds* (one bud at each node), *opposite buds* (two at each node), *whorled buds* (three or more at each node).

A typical bud consists of a mass of meristematic tissue, which by cell division, enlargement, and maturation results in growth in length of the stem and the production of leaves or flowers. An examination of a longitudinal section of a stem tip shows a meristematic region in the bud, a region of elongation, and a maturation region. These regions usually have a greater longitudinal extent in stems than they do in roots. Terminal buds increase the length of the twigs at whose apices they are situated. Lateral buds usually form branches of the twigs on which they are located, together with leaves or flowers or both. Thus the branches of a stem have an *external* origin, as compared with the internal origin of branch roots.

In most leaf buds, there are small lateral protuberances of the meristematic tissue; these protuberances develop with the growth of the bud into leaves. A bud is thus a much shortened twig, with nodes, internodes, and leaf rudiments. In naked buds, the meristematic tissue is uncovered; in covered buds, it is covered by overlapping scales which protect the meristematic tissue against temperature extremes, desiccation, and often against the entry of parasites. These scales are often tough, and thick, and sometimes have waxy or sticky secretions which resist water loss.

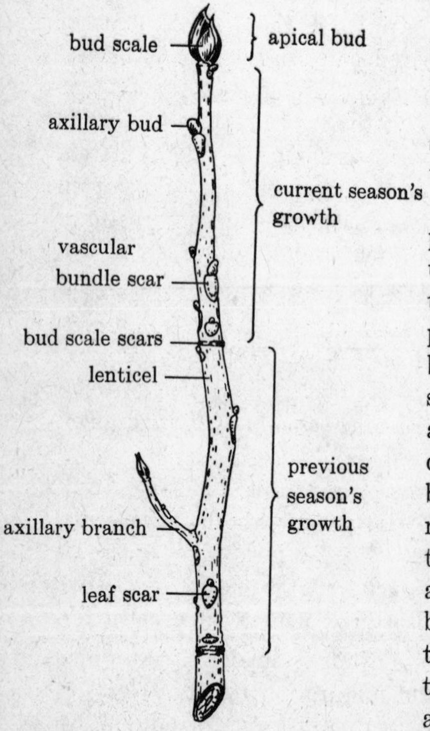

In annuals and in woody plants in the tropics, the buds usually grow continually during the life of the plant. In woody plants in temperate zones, the buds grow during a growing season, but remain dormant during a season of low temperature or of scant moisture. A bud grows by the elongation of the internodes which it contains and by the formation of new cells at its apex. In covered buds, the scales bend outward and fall away as the internodes of the bud begin to elongate. As the internodes and tiny leaves of a bud develop, axillary buds are formed in the axils of the newly formed leaves,

Fig. 7/1. Two-year-old twig of hickory.

bud scale — apical bud
axillary bud
current season's growth
vascular bundle scar
bud scale scars
lenticel
previous season's growth
axillary branch
leaf scar

and toward the end of the growing season, a new terminal bud is formed. In temperate zone woody plants, these buds, formed in one growing season but destined to remain dormant through a winter before the next growing season, are called *winter buds.* The general form of a shoot system depends upon the position, number, kinds, and degree of activity of its buds.

Other Structures. Herbaceous stems bear relatively few structures besides leaves and buds. There may be present *hairs,* which are outgrowths of epidermal cells, and *spines,* which may be modified twigs, leaves, hairs, or stipules.

Woody stems show a greater variety of structures than herbaceous stems, particularly during winter when they usually drop their leaves. Among the structures, in addition to leaves and buds, on woody twigs (Fig. 7/1) are:

LENTICELS.   Tiny raised pores for gaseous exchange.

LEAF SCARS.   Usually crescent-shaped or circular marks left by the fall of leaves. They are the places at which the leaf stalks grew from the stem.

BUNDLE SCARS.   Appear as tiny raised dots in the leaf scars. These are the broken ends of *vascular bundles* (conducting strands of xylem and phloem) which extend from the conducting tissues of the twig into the leaf stalk.

BUD SCARS.   Ring of small narrow scars left by the falling away of the bud scales and forming a complete thin circle around a twig. Each of these rings of bud-scale scars marks the place at which a terminal bud began its growth. Since a new terminal bud is developed each year, the number of bud scars on a twig indicates the age of the twig.

TWIG SCARS.   More or less circular scars left by the falling away of branch twigs, a natural phenomenon in many plants. Similar to twig scars are *fruit scars,* left by the falling away of fruit stalks.

As woody stems grow in diameter, the smooth-growing bark is split and is replaced by rough bark. With the disappearance of the smooth bark, the lenticels and scars described above disappear. In some woody plants, the bark remains smooth for many years, and some of these structures can still be seen on old branches and trunks, as in white birch trees.

SPECIALIZED STEMS

These are stems which have more or less unusual or specialized functions. Among the common types of specialized aerial and subterranean stems are:

Specialized Aerial Stems.

TENDRILS of grape and Boston-ivy. Climbing organs.

THORNS of honey locust and certain other species. Offer protection.

STORAGE STEMS of cacti. Store food and water and are photosynthetic.

AERIAL BULBS of onions. Serve for reproduction.

RUNNERS (STOLONS) of strawberry and other plants. Produce new plants at their nodes when these touch the ground.

CLADOPHYLLS of *Ruscus* (butcher's broom). Photosynthetic, leaflike expansions of stems, with true leaves that are temporary and non-functional.

Specialized Subterranean Stems.

RHIZOMES.   A rhizome is a horizontal stem growing at or beneath the surface of the ground. They may be slender (quack grass) and serve chiefly for reproduction, or may be swollen by stored food (iris). Rhizomes are perennial and send up new shoots year after year from their buds.

TUBERS.   A tuber is an enlarged tip of a rhizome. It stores much food and is located at the end of a slender rhizome (Irish potato). The "eyes" of a potato are buds which are capable of growing into aerial stems. Pieces of tubers with one or two eyes are used to propagate Irish potatoes.

BULBS.   A bulb is a large, rather globose perennial bud with a small basal stem at its lower end, from which grow fleshy, scale-like, overlapping leaves (onion, tulip). Bulbs serve chiefly for storage, reproduction, and for carrying the plant through seasons unfavorable for active growth.

CORMS.   A corm is a rather globose perennial stem with thin, papery leaves on its surface (gladiolus, crocus). Most of a corm is stem, in contrast to a bulb, most of which consists of storage leaves. Corms serve for storage, reproduction, and for carrying the plant through an unfavorable season.

Chapter 8

The Internal Structure of Stems

As stated in the preceding chapter, stems are usually divided into two groups: *woody* and *herbaceous*. Neither the woody nor the herbaceous categories are homogeneous but include many variations in anatomical features.

It is generally believed that woody stems are more primitive than herbaceous ones and that herbaceous stems have developed in the course of evolution from woody types, as a result of progressive cooling and drying of the earth, beginning at the poles and extending toward the equator.

THE INTERNAL STRUCTURE AND DEVELOPMENT OF WOODY STEMS

Gross Internal Structure. An examination, with the naked eye, of a cross-section of a woody stem (Fig. 8/1) shows two major groups of tissues: the *bark*, which forms the outer part of the stem, and the *wood*, or *xylem*, which forms the inner part of the stem. Inside the xylem in some kinds of woody stems a parenchymatous tissue, called *pith*, is present. Between the bark and the wood is the *cambium*, a layer of meristematic cells which by cell division cause an increase in diameter of the stem. In the outer part of the bark there develops a *cork cambium*, which forms *cork cells* in the outer part of the bark. The cambium forms wood cells much more rapidly than it does inner-bark cells; thus, as stems grow older the proportion of wood to bark increases.

Microscopic Internal Structure. Under low magnification, additional stem structures become apparent. We place these tissues in two broad categories—the *primary tissues* and the *secondary tissues*.

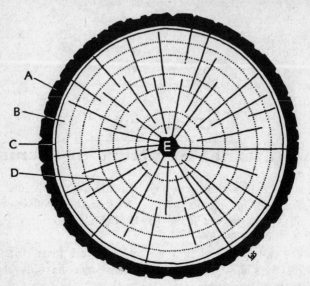

Fig. 8/1.   Cross-section of woody stem. A. Bark. B. Wood, showing annual rings. C. Cambium. D. Vascular ray. E. Pith. (about 4 ×)

PRIMARY TISSUES OF WOODY STEMS. (Fig. 8/2). Primary tissues are those formed by the growth of terminal meristems, which, in stems, are situated in buds. In woody stems, the primary tissues are formed in the early part of the first season's growth of the twig, but as a twig grows after its first year, the newly formed tissues are entirely *secondary*—that is, they develop from the cambium and cork cambium. The primary tissues of a woody stem are the *epidermis*, the *cortex*, and the *stelar* (*fibro-vascular*) *tissues.*

The epidermis. This is a single surface layer of cells, usually with cutinized, nearly waterproof outer walls. The epidermis is a protective tissue, the chief value of which lies in its preventing excessive evaporation from the tissues inside.

The cortex. This area varies in thickness in different kinds of stems. It usually contains strengthening *collenchyma* cells, storage *parenchyma* cells, and frequently strengthening *fiber* or *stone* cells. The cortex is thus a region of protection, strength, and storage.

The stelar (*fibro-vascular*) *tissues.* The tissues of the stele are the *pericycle, phloem, cambium, xylem, rays,* and *pith.* The *pericycle* consists usually of parenchyma cells and often of strengthening fiber cells. These fiber cells in some plants (flax, hemp) are important in the manufacture of thread, twine, and textiles. The *phloem* tissue is just inside the pericycle and consists of sieve tubes, companion cells, fibers, and parenchyma cells. The sieve tubes and companion cells conduct foods downward, the fibers provide strength, and the parenchyma cells store various substances. The *cambium* is a usually continuous layer of meristematic cells just inside the phloem. By their division, the cambium cells form new xylem cells inside themselves, new phloem cells outside themselves, and thus cause the stem to grow in diameter. The *xylem* consists of *vessels, xylem fibers, tracheids,* and *xylem parenchyma* cells. The vessels and tracheids serve primarily for conduction, secondarily for support. Xylem fibers are thick-walled strengthening cells. Parenchyma cells store foods and other substances. In the wood (xylem) of some species of plants, one or more kinds of these cells may be absent; for example, the wood of most gymnosperms lacks vessels and xylem fibers. Sometimes other types of cells occur in xylem, e.g., *resin canals* in the wood of certain gymnosperms. A *ray* is a band of cells (chiefly parenchyma) which extends radially in a stem, that is, along a radius of the stem. Rays function chiefly for transverse conduction of materials in stems and for food storage.

The epidermis and cortex are continuous layers of cells forming what appear in cross-section to be complete circles in the stem. The primary xylem and primary phloem of some species are likewise continuous layers; in other species, they occur in separate strands, or *vascular bundles.* The cambium of woody stems is usually in the form of a complete circle of cells between the xylem and phloem tissues. In virtually all seed plants the xylem tissue is inside the phloem, with the cambium between xylem and phloem.

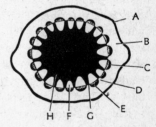

Fig. 8/2. Primary tissues of woody stem. A. Epidermis. B. Cortex. C. Pericycle fibers. D. Pericycle parenchyma. E. Primary phloem. F. Primary xylem. G. Cambium. H. Pith. (about 10 ×)

SECONDARY TISSUES OF WOODY STEMS.   These tissues comprise the *cambium* and the *cork cambium* and their derivatives. (Fig. 8/3.)

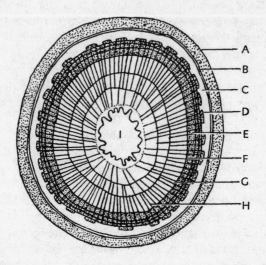

Fig. 8/3. Cross-section of three-year-old woody stem. A. Cork. B. Cork cambium. C. Cortex. D. Pericycle fibers. E. Secondary phloem. F. Cambium. G. Secondary xylem (three annual rings). H. Vascular ray. I. Pith.

The Cambium and Its Derivatives.   After primary tissues of a stem are formed, little or no further growth in length of the particular stem segment occurs. These primary tissues of woody stems are formed in the first season's growth. Also, in the first season's existence of a particular stem segment, cambial activity leads to secondary growth. From this time on, all growth in this part of the twig is secondary. Since the cambium is outside the xylem, the secondary xylem cells are produced outside the primary xylem cells, which thus remain the innermost xylem cells of a stem. The most recently-formed xylem cells (secondary xylem) are outermost, just inside the cambium. The new (secondary) phloem cells are produced by the cambium just outside itself; the outermost phloem cells of a stem are primary cells and are thus the oldest of the phloem. In most woody stems, the cambium is active along its entire length, thus forming continuous cylindrical layers of xylem and phloem cells. These continuous circular layers are more noticeable in the xylem than in the phloem, because more secondary xylem is usually produced than secondary phloem. The layer of xylem formed by one year's growth activity of the cambium is called an *annual ring*. The kinds of cells formed in secondary xylem and phloem tissues are the same kinds found in primary tissues: namely, tracheids, vessels, etc., in the xylem; and sieve tubes, companion cells, etc., in the phloem. The cambium likewise forms *secondary rays* in its growth activity and also increases the radial length of the *primary rays*. Primary rays are those which extend inward to the pith. Secondary rays never extend as far toward the center of the pith as the primary rays. As the cambium grows, it is gradually forced outward by the abundant secondary xylem which it has produced inside itself. Cell division occurs radially as well as tangentially in the cambium cells and thus the circumference of the cambium is increased as it is pushed outward by the secondary xylem.

The Cork Cambium and Its Derivatives. The *cork cambium* is a secondary meristematic tissue which develops from certain parenchyma cells in the outer part of the cortex. The cork cambium produces cells on both its inner and outer faces, chiefly the latter. The cells formed on the outer face of the cork cambium are *cork cells*. The walls of these cells become *suberized* and thus waterproof and the cork cells die. The accumulated cork cells form the rough, hard outer bark of woody stems; the outer bark reduces evaporation of water from the tissues inside it, protects the inner cells against extremes of temperature, and against mechanical injury. In addition to the cork, cork cambium, pericycle, and cortex, the bark also contains the phloem (inner bark); thus, one of the functions of bark is food conduction. The thickness of bark varies on stems of different ages and in different species of trees. The lenticels in young bark serve for exchange of gases—chiefly the exit of carbon dioxide and the entry of oxygen. As stems grow older, the outer layers of cork are split lengthwise by the more rapidly-growing wood within, and the lenticels are no longer apparent.

The Structure of Wood. As stated in the preceding section, wood is xylem tissue and is made up (in angiosperms) of wood fibers, vessels, tracheids, and wood parenchyma. In gymnosperms, the wood is simpler in structure, being composed almost entirely of tracheids. Angiosperm woods are hardwoods, gymnosperm woods are softwoods.

Wood is composed chiefly of *cellulose* and *lignin,* a compound often associated with cellulose. Also in wood are water, pigments, minerals, oils, starch, gums, etc.

ANNUAL RINGS. In the temperate zones and in parts of the tropics in which there is a definite alternation of growing and dormant seasons, the secondary xylem tissue is produced in layers (*annual rings*) by the cambium. Usually one annual ring is formed each year; thus, the age of a tree or branch may be determined by counting the number of anual rings seen in the cross section. An annual ring (Fig. 8/4) is composed typically of two somewhat distinct bands of cells; the inner band of each ring is made up of rather large cells formed by the cambium in the

Fig. 8/4. Pine Wood. A. Summerwood of annual ring. B. Springwood of annual ring. C. Side view of vascular ray. D. End (tangential) view of vascular ray. E. Tracheids. F. Cambium. G. Phloem.

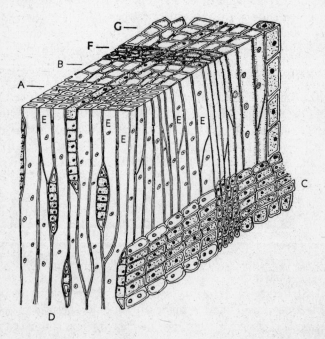

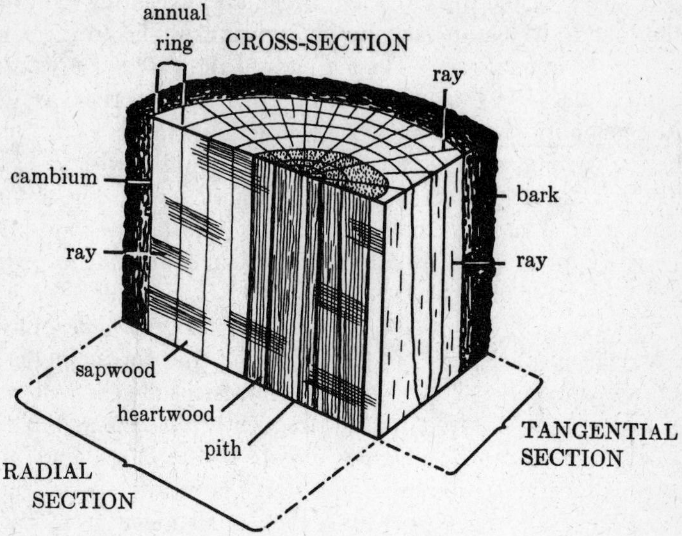

Fig. 8/5. Diagram of a section of a woody stem, cut along three planes.

spring and hence called *springwood;* the outer band of each ring is made up of smaller, and sometimes thicker-walled cells, formed in the summer and hence called *summerwood.* The innermost annual ring in a tree is the oldest, the outermost (just inside the cambium) the youngest. The widths of annual rings vary greatly; these variations are reflections of differences chiefly in rainfall, but also in soil aeration and in light which obtained while the ring was being formed. Thus, it is possible to trace climatic cycles by studying the annual rings of trees.

WOOD SECTIONS. Sections can be cut along any one of three planes:

1. Those cut *across* the longitudinal axis of the stem. In such a *transverse,* or *cross* section, the annual rings appear as concentric circles, with the vascular rays radiating out across the rings, like the spokes of a wheel.

2. Those cut lengthwise through the stem. These are of two types—*tangential* sections and *radial* sections. Those cut tangentially to the annual rings are at right angles to the radiating vas-

cular rays. In such a tangential, or *slab* section, the annual rings appear to the naked eye as broad, irregular, alternating light and darker bands, which are the alternating summer and spring wood bands of the annual rings. The cut ends of the vascular rays may be seen as flecks in the annual rings. Tangential sections are the most commonly seen sections in construction lumber.

Radial sections are cut through the pith, parallel with the vascular rays. In such a radial or *quarter-sawed* section, the annual rings appear as narrow, longitudinal, light and dark bands, with the rays appearing as usually smooth bands running at right angles to the annual rings.

HEARTWOOD AND SAPWOOD. As woody stems grow older, physical and chemical changes occur in the wood. The conducting cells of the xylem become plugged up with protrusions of cells (*tyloses*) into their cavities. Also, there is frequently an increase in the amount of *tannins* (bitter substances), *resins, gums,* and pigments in wood cells as they grow older. The young annual rings are not plugged and usually have but small amounts of these chemical compounds; these rings conduct actively and have numerous living cells. These rings constitute the *sapwood* of a tree. The older, plugged rings with larger quantities of tannins, etc., and with no living cells constitute the *heartwood,* which is usually darker in color than sapwood. The sapwood lies outside the heartwood. The plugged heartwood cells do not conduct materials. Heartwood is much more valuable than sapwood for outdoor construction because it decays much less easily, and also for indoor use because its colors are attractive. The resins, tannins, etc., are poisonous or distasteful to wood-rotting organisms, and the plugged cells of heartwood make the entry of organisms and moisture more difficult. The relative proportions of heartwood and sapwood vary in different kinds of trees.

THE INTERNAL STRUCTURE AND DEVELOPMENT OF HERBACEOUS STEMS

All gymnosperms have woody stems. Some angiosperms have woody stems, others have herbaceous stems. Of the angiosperms, the monocotyledons are almost entirely herbaceous, whereas the dicotyledons include both herbaceous and woody forms. In herbaceous stems generally secondary growth is lacking or is relatively slight, as compared with that of woody stems. The xylem and phloem tissues of herbaceous stems contain much the same kinds of cells as those of woody stems.

Herbaceous Stems of Dicotyledons (e.g., bean, sunflower). As it grows, the terminal bud forms epidermis, cortex, and stele, as in the primary growth of woody stems. The epidermis is structurally and functionally similar to that of woody stems; the cortex is often thinner than that of woody stems but is quite similar; the stele is composed of primary phloem, cambium, primary xylem, and pith, as in a young woody twig (Fig. 8/6).

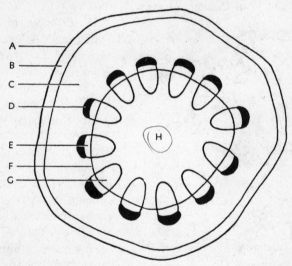

Fig. 8/6. Cross-section of herbaceous dicot stem. A. Epidermis. B. Sclerenchyma tissue. C. Cortical parenchyma. D. Pericycle fibers. E. Phloem. F. Cambium. G. Xylem. H. Pith.

There is a much larger proportion of pith and cortex than of xylem and phloem in herbaceous stems as compared with woody stems. Pericycle fibers are sometimes present. In some plants, the xylem and phloem are in distinct vascular bundles; in others, they are more or less continuous. The vascular bundles in dicots are usually arranged in a single circle.

The cambium is located between the xylem and phloem, with the xylem inside the cambium and the phloem outside. In some species, the cambium occurs only in the vascular bundles, in other species, it is a continuous layer. In either case, the cambium usually shows relatively little activity and there is thus little secondary growth in most herbaceous dicots.

Herbaceous stems are chiefly annual, but in some cases may live for more than one year, in which case considerable secondary growth occurs and the stem becomes more or less woody.

Herbaceous Stems of Monocotyledons (e.g., corn). Most monocot species (exceptions are palms and a few others) have no cambium and thus no secondary growth. Such stems grow relatively little in diameter; the little diametric growth which does occur results from the enlargement of primary cells (Fig. 3/2).

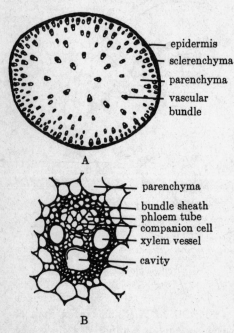

epidermis

sclerenchyma

parenchyma

vascular bundle

A

parenchyma

bundle sheath
phloem tube
companion cell
xylem vessel

cavity

B

Fig. 8/7. A. Diagram of a cross-section of a monocot stem (corn). (Magnification about 20 ×.) B. Cross-section of a single vascular bundle. (400 ×.)

In all monocots, the xylem and phloem are arranged in vascular bundles, with the xylem in the inner part of the bundle, the phloem toward the outer part. The vascular bundles of monocots are not arranged in a circle as in dicots, but are scattered through the parenchyma tissue which usually fills the stem. The lack of cambium and the scattered bundles are outstanding features of monocot stems (Fig. 8/7).

The surface of most monocot stems is covered by an epidermis similar in structure and function to that of a dicot stem. There is no definite demarcation between pith and cortex; parenchyma cells are continuous throughout the stem, surrounding the vascular bundles.

The vascular bundles of monocot stems are surrounded by a layer of thick-walled strengthening cells. Such bundles without cambium are called *closed bundles,* as compared with the *open bundles* of dicot stems.

Leaf Traces and Leaf Gaps. At each stem node, vascular bundles separate from the xylem and phloem tissues of the stem, pass out through the cortex and enter the leaf-stalk. These vascular bundles are called *leaf traces.* Wherever a leaf trace departs from the vascular tissues of a stem, a small break occurs. Such a break is called a *leaf gap.* (Fig. 3/1.)

Leaf traces constitute the connections between the veins (vascular tissues) of leaves and the vascular tissues of stems.

Chapter 9

The Physiology of Stems

The principal functions of stems are the conduction of materials upwards, downwards, and transversely, and the production and support of leaves and reproductive structures. The production of leaves, internodes, and reproductive structures is the function of buds. The function of support is performed by the strengthening cells which occur in the pericycle, xylem, and phloem, as described in the preceding chapter.

Other more specialized functions of stems are food storage, climbing, reproduction, water storage, etc., as described in an earlier chapter.

The physiologically most complex function of stems is conduction, or translocation.

CONDUCTION OF MATERIALS BY STEMS

Conduction by Xylem. The chief function of xylem is the upward conduction of water and dissolved substances (*sap*). These dissolved substances are chiefly mineral salts from the soil; also, foods previously stored in the roots and stem may be conducted upward in the xylem, especially in spring in woody plants. Evidences for upward conduction in the xylem are:

1. A ring of bark (of which the phloem is a part) may be removed from a stem, with no interference with the normal condition of the leaves. If the xylem be severed but the phloem left intact, wilting of the leaves follows, for their supply of water and solutes has been shut off.

2. If the cut lower end of a stem is allowed to stand in a water solution of a dye for several minutes or hours and if cross-sections of the stem are then cut and examined, the dye will be

found principally in the cells of the xylem. In some plants, there may be a small amount of upward conduction through the phloem.

Enormous quantities of water move upward through stems. For example, a single corn plant loses by evaporation about 50 gallons of water during its few weeks of life, a sunflower plant evaporates about 150 pounds of water in its life span of about 140 days. These figures represent only the water which evaporates from plants. In addition, considerable amounts of water and minerals rise to the leaves and are made into foods which remain in the plant.

The exact nature of the forces responsible for the rise of these materials is not known, but several explanations of the *ascent of sap* have been proposed, as follows:

ROOT PRESSURE. This is responsible for the exudation of sap from cut stems, as from pruned grapes in early spring. The magnitude of root pressure rarely exceeds two atmospheres, except under special laboratory conditions (one atmosphere will push water up only 30 feet). Root pressure usually is demonstrable in nature only in early spring before leaves have developed and when upward sap movement is rather slow. During the growing season when leaves are full grown and upward sap movement is most rapid, root pressure is usually nil. Thus, root pressure is probably a negligible factor in sap rise in large trees during summer.

CAPILLARY ATTRACTION. An example is the rise of liquids in tubes of very small diameter. Capillarity acts only in open-ended tubes, and since xylem tubes are closed at both ends, it is probably a negligible factor in sap rise.

ATMOSPHERIC PRESSURE. It has been suggested that atmospheric pressure forces water up through the conducting cells of the xylem, as it supports a mercury column in a barometer. Under ideal conditions, atmospheric pressure could not support a water column higher than 33 feet at sea level. Thus, it could not be more than a very minor factor in sap rise, even if some force were known to be at work producing the necessary negative pressure at the top of the plant.

ACTION OF LIVING CELLS. It has been suggested that living cells adjoining the conduction cells of the xylem force water upward, possibly by some sort of pumping action. However, experiments have shown that dead xylem can conduct sap very readily. Thus, the action of living cells is probably of secondary importance in the rise of sap, and has, moreover, never been experimentally demonstrated.

IMBIBITION IN CELL WALLS OF XYLEM.   The cell walls of xylem cells absorb water by imbibition, but imbibitional force is not sufficient to explain either the rapidity or magnitude of sap rise. Further, the greater part of the sap moves through cell cavities, not cell walls.

TRANSPIRATION PULL AND WATER COHESION.   *Transpiration* (evaporation) from leaves causes a great water deficit in the leaf cells. As a result of this deficit, water is drawn osmotically from the xylem cells in leaf veins by the cells surrounding the veins. Thus, a pull is exerted in the uppermost xylem cells of plants, those in the leaves. Water molecules have tremendous cohesive power (ie., they remain together with great mutual attraction), often as much as 150 atmospheres. The pull exerted by the movement of water from the uppermost xylem cells of leaves into the partially dried leaf cells is transmitted through the water columns in the xylem of stems down into the roots. Thus, water is *pulled* up through stems by the evaporation-pull of transpiration and the cohesive power of water molecules. The osmotic pressures in transpiring leaf cells frequently reach 30 atmospheres, more than enough to cause the rise of water to the tops of the highest known trees. (Twenty atmospheres would be sufficient to cause the ascent of sap to the highest known trees.) This attempt to explain sap rise is accepted by most botanists. It is likely that such factors as imbibition, root pressure, the action of living cells, etc., are contributory factors in sap rise.

Conduction by Phloem.   The chief function of phloem is the downward conduction of foods manufactured in the leaves. In some cases, some upward conduction of minerals and foods occurs in the phloem. Evidences for downward conduction of foods in the phloem are:

1. Chemical analyses show the presence of larger amounts of foods in the phloem than in the conducting cells of the xylem, and the use of radioactive tracers shows the exact paths taken by moving substances.

2. If a complete girdle of bark (including phloem) is removed

from the stem, an enlargement develops immediately above the girdle. This swelling is caused by the accumulation of foods above the girdle, since they are unable to pass across the gap and continue their downward journey.

The forces responsible for the passage of foods through the phloem are not well understood. The rate of such movement is usually too rapid to be explained by simple diffusion. The pores in the end walls (*sieve plates*) of sieve tubes facilitate such movement.

Conduction by Vascular Rays. The vascular rays which extend radially in stems extend through xylem and also into the phloem. These rays bring about transverse (crosswise) conduction of foods, minerals, water, and gases. Oxygen entering the lenticels finds its way into the outer ends of vascular rays, some of which terminate just under the lenticels, and is conducted by ray cells into the inner tissues of the stem. Similarly, carbon dioxide is conducted outward to the lenticels from which it passes out into the air.

OTHER FUNCTIONS OF STEMS

Food Storage. Storage occurs chiefly in the parenchyma cells of cortex, rays, and pith, and also in the phloem and xylem parenchyma. Water and salt storage also occur in these tissues. In specialized stems, such as tubers, corms, etc., these storage tissues are extensive. Also stored in certain types of stems are *resins*, *gums*, *latex* (a milky fluid), etc.

Climbing Stems. Tendrils, such as those of grapes, or twining stems, such as those of morning-glories, are examples.

Food Manufacture. Stems which contain chlorophyll, such as those of most herbaceous plants and of cacti, manufacture food. In the absence of leaves, the stems of these plants also store water and food.

Reproduction. In stems, reproduction is brought about by runners, by the branching of rhizomes, by the formation of new bulbs and corms from older ones, etc.

PRACTICAL APPLICATIONS OF A KNOWLEDGE OF STEM STRUCTURE AND PHYSIOLOGY

A knowledge of the structure and function of stems is fundamental in the horticultural practices of *pruning, girdling,* and *grafting.*

Pruning.   In pruning away a diseased, broken, or otherwise undesired branch, the cut should be made as close to the main branch as possible and parallel to it, in order that the growth tissues surrounding the wound may form new tissues to heal the wound. Since the food moving down through the phloem comes from leaves above the wound, the wound must be in a position near this food, if healing is to occur. If the cut is made some distance from the main branch and a stump is thus left, healing of the wound at the end of the stump does not occur because there are no leaves above it—these were removed when the branch, of which the stump is a remnant, was cut off. A pruning wound should be painted to prevent the entry of molds, bacteria, etc., which might cause the wood to rot.

Girdling.   Girdling is the removal of a complete ring of bark (including the phloem). Girdling stops the downward passage of food, which collects above the girdle. Girdling is practiced in order to: (1) Produce large fruits. The accumulation of food causes the fruits above a girdle to become abnormally large. (2) Kill trees, in clearing land. A girdle on the main trunk prevents the downward passage of food and thus starves the roots. (3) Produce more flowers in the season following girdling.

Grafting.   Grafting is a horticultural practice in which two freshly-cut stem surfaces are bound together in such fashion that their cells grow together; thus there is formed a union between the two pieces of stem. The basal, rooted stem used in a graft is the *stock;* the stem piece which is grafted to the stock is the *scion.* Stock and scions are prepared in various ways which vary according to their size, their internal anatomy, etc. A common type of grafting is *budding,* in which the *scion* consists of a single bud which is placed in a slit in the bark of a young stock.

Precautions necessary to ensure a successful graft are: (1) The stock and scion must be placed together in such a manner that their cambial layers are in contact, for the cambium produces the growth which causes union. (2) Stock and scion must be firmly bound together so that the cut ends of the stem pieces do not move in the wind. (3) The region of the graft should be covered with wax to prevent the entry of parasites into the cut and to prevent the drying out of the exposed tissues. (4) Successful grafts can be made only when stock and scion are very closely related, for only in closely related species are the growth habits, anatomy, and biochemical processes enough alike to ensure union.

Grafting is employed for several purposes, some of which are: (1) to propagate seedless varieties; (2) to propagate hybrids, the seed of which produce offspring of different kinds; (3) to propagate plants the seeds of which germinate poorly; (4) to produce more rapid fruiting, since some grafted fruit scions produce fruit sooner than saplings raised from seed; (5) to change or fashion the shape of a plant, as in the umbrella catalpa; (6) to check or eliminate parasites which damage the roots of the variety from which the scion is taken but to which the variety of stock on which the scion is grafted is immune; (7) to acclimate certain plants to environments in which their tops grow but which are unfavorable to their roots.

Grafting never produces new kinds of plants. It is exclusively a *vegetative* mode of propagation, and stock and scion always maintain their individualities—there is never any mixing of characters in a genetic sense, although each member of a graft may affect the other just as any environmental factor might. There are some apple trees in which over a hundred scions have been grafted to the same stock. Each scion continues to produce its own variety of apples.

Chapter 10

The Leaf:
Structure and Physiology

A leaf is a lateral outgrowth of a stem, arising at a node, and possessing a bud in its axil. Most leaves are flattened and expanded, but modified or specialized kinds of leaves do not necessarily have this flattened structure. The common foliage leaf is primarily a food-manufacturing organ.

ORIGIN AND ARRANGEMENT

Leaves develop from the growth (meristematic) tissues of buds. They arise as lateral protuberances of the stem apex, and as the bud grows and expands, the protuberances enlarge and become differentiated into leaves. Leaves may be arranged on a stem in three ways (Fig. 10/1).

Spiral or **alternate.**   In this kind of arrangement, a single leaf is produced at a node: e.g., elm, apple, oak.

Opposite.   Here, there are two leaves at a node, usually directly opposite each other: e.g., ash, maple.

Whorled.   In a whorled arrangement three or more leaves grow at the same node: e.g., catalpa.

LONGEVITY OF LEAVES

In most of the plants in the temperate zones, and in the tropical regions where the rainfall is seasonal, the leaves live for only a single growing season, then they fall off. Such plants are known as being *deciduous*. Plants which retain their leaves for more than one growing season and thus have living leaves at all times of the year are called *evergreens*. The leaves of most evergreens live but one calendar year, although a few may persist more than four years.

EXTERNAL STRUCTURE OF FOLIAGE LEAVES

A typical foliage leaf consists of a stalk, or *petiole,* which grows out from a node, and an expanded portion, the *blade.* In many leaves, in addition, small flaps of tissue, the *stipules,* grow out from the base of the petiole. Stipules may be small and apparently functionless or large and leaflike, as in Japanese quince. *Sessile* leaves lack petioles. Plants with photosynthetic stems (as cactus and asparagus), or chlorophyll-less parasites (as dodder), may have small, non-green, scalelike, ephemeral leaves.

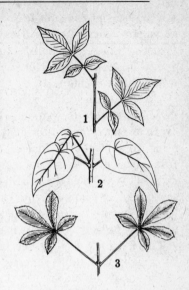

Fig. 10/1. Variation in arrangement and morphology of leaves. 1. Alternate, pinnately compound leaves with serrate margins (hickory). 2. Whorled, simple leaves with pinnate venation and entire margins (catalpa). 3. Opposite, palmately compound leaves with serrate margins (horsechestnut).

The Leaf Blade. The flat, thin blades of leaves are adapted for the easy penetration of light and carbon dioxide. Variations in leaf morphology are commonly used in plant identification.

LEAF SIZE. Leaves vary from a fraction of an inch in *Wolffia* to ten or fifteen feet in banana or even to fifty feet in some palms.

LEAF SHAPE. Leaf blades vary in shape from *linear* (grasslike or needle-shaped) to almost *circular* in nasturtium, with many gradations of form between these extremes.

LEAF VENATION (arrangement of veins). Two types of venation are *parallel* and *net.* In parallel venation, the veins run alongside each other from the base of the blade to the tip (iris and corn), or alongside each other and at a similar angle to the *midrib* of the blade (banana and canna). In net venation, the veins branch out repeatedly and form a network over the blade. Two types of net-veined leaves are *pinnate,* which has one midrib (elm, oak), and *palmate,* in which several large veins of equal size branch into the blade from the end of the petiole (maple, geranium). Parallel-veined leaves are characteristic of monocotyledons (grasses, cattails, lilies, bananas, orchids, etc.); netveined leaves are characteristic of dicotyledons (roses, oaks, willows, sunflowers, etc.).

LEAF MARGINS. The edges of leaves may be *entire* (smooth), variously *serrate* (saw-toothed), or *lobed*. Lobes may be pinnate (with one main midrib) or palmate (with several main veins radiating from the base of the leaf). If the leaf with its petiole, coming from a single node, has one blade, it is *simple*. If the lobing is so deep as to extend to the midrib or petiole, then one leaf may have several subdivisions (*leaflets*) and the leaf is *compound*. A leaflet may in turn be divided, making a *twice compound* leaf. Some ferns may be compounded as many as five times.

The Petiole. The petiole is usually more or less cylindrical, with vascular bundles running lengthwise through it, connecting the veins of the blade with the xylem and phloem of the stem. In some monocots (grasses), the petiole has the form of a sheath which clasps the stem (Fig. 3/2). Petioles bend actively in response to the stimuli of light and gravity, and are able to hold the blade advantageously with reference to the direction of light.

Fig. 10/2. Cross-section of leaf. A. Cutin layer (cuticle). B. Upper epidermis. C. Palisade tissue of mesophyll. D. Spongy tissue of mesophyll. E. Xylem of vein. F. Phloem of vein. G. Air spaces. H. Lower epidermis. I. Guard cells. J. Stoma.

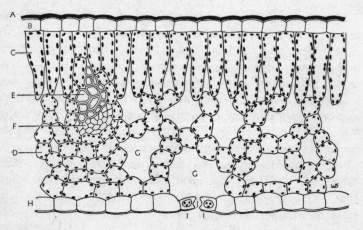

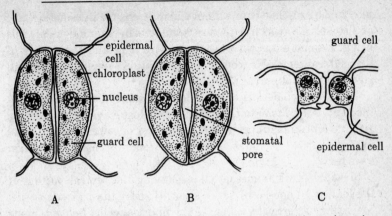

Fig. 10/3. A. Surface view of a closed stoma. B. Surface view of an open stoma. C. Section through a closed stoma, showing differential thickening of guard cell walls. (Clover, magnification 1500 ×.)

INTERNAL STRUCTURE OF A FOLIAGE LEAF

A cross-section of a leaf under the microscope shows three regions: *epidermis* (upper and lower); *mesophyll;* and *veins* (Fig. 10/2).

Epidermis.   This is usually a single layer of cells on the upper and lower surfaces of the leaf. The cells fit tightly together and are of two types: ordinary epidermal cells, and crescent-shaped *guard cells* which contain chloroplasts and occur in pairs, with an opening or *stoma* enclosed by each pair. The ordinary epidermal cells protect the inner tissues from desiccation, from mechanical injury, and to some extent from the entrance of parasites. They often secrete a waxy *cutin* on their outer surfaces. The waterproof cutin layer, or *cuticle,* varies in thickness in different species and is effective in reducing water loss.

The stomata are avenues of exchange of carbon dioxide, oxygen, and water vapor between the interior of the leaf and the external atmosphere. The opening and closing of stomata is related to the action of light. The thickened inner walls and relatively thinner outer walls of guard cells make possible the changes in shape which result from changes in water content (Fig. 10/3 b, c). The immediate cause of changes in guard cell shape is change in water content, turgid guard cells making open stomata, and flaccid ones making the stomata close. The most likely cause of change in water content is change in osmotic value of the cell sap, which is in turn regulated by solute content, this being in some unknown way affected by light and the acidity of the cell contents. No completely acceptable explanation of all guard cell action is available.

The upper epidermis of most leaves usually has a thicker cuticle and fewer stomata than the lower. Most leaves have between 100 and 350 stomata per square millimeter of lower surface, making a mature, moderate-sized leaf contain easily a million stomata.

Mesophyll. The mesophyll, occupying the central portion of the leaf, is composed of two distinct tissues: the *palisade* tissue, consisting of vertically elongated, cylindrical cells. Below the one or two *palisade* layers is the *spongy* tissue, composed of loosely-packed cells of variable form. Both layers are rich in chlorophyll and constitute the food-making tissues of the leaf. The numerous intercellular spaces in these tissues make possible the ready diffusion of gases to all cells.

Veins (Vascular Bundles). Veins are branched continuations in the mesophyll of the vascular bundles of the petiole. A vein consists usually of xylem cells (vessels, tracheids) and phloem cells (sieve tubes and companion cells), which conduct, respectively, water and mineral salts upward into the mesophyll, and foods downward into the petiole. In most leaves, the xylem cells are in the upper part of the veins, the phloem cells on the lower. Larger veins are surrounded by thick-walled strengthening cells which aid in supporting the blade.

THE PHYSIOLOGY OF LEAVES

Photosynthesis. Photosynthesis entails the production of complex organic compounds from carbon dioxide and water, energized by light and utilizing the chlorophylls, accessory pigments, and associated enzymes. It is the basic process of energy storage in nature and all animals and plants (except a few bacteria) depend upon it. It is also the only important source of atmospheric oxygen.

Several conditions are necessary for photosynthesis:

CARBON DIOXIDE. Carbon dioxide makes up 0.03—0.04% of air; it reaches the interior of the leaf through the stomata and enters the mesophyll cells dissolved in water. Carbon dioxide concentration is frequently the "limiting factor" of photosynthesis because, with other factors being kept constant, an increase in CO_2 concentration can raise the photosynthetic rate. In winter, temperature can be limiting; or in deep forests or at night, light can be limiting. Any essential factor (e.g., chlorophyll content, water supply, etc.) can, if it is the one keeping the photosynthetic rate down, be a limiting factor. An increase in the carbon dioxide content of the atmosphere up to 0.9% can increase proportionately the rate of photosynthesis in some plants.

WATER.   Water is absorbed by the roots and passes through the conducting tissues of the plant to the mesophyll cells. Only about 1% of the water taken in is converted by photosynthesis into other compounds.

TEMPERATURES.   The favorable temperature for photosynthesis to occur is usually between 5°C. and 45°C.

LIGHT.   Any wavelength of visible light may be photosynthetically effective. Chlorophyll extracted from chloroplasts and observed in solution (e.g., in acetone) absorbs selectively in the blue and red portions of the spectrum of white light, but intact leaves are less selective, and a leaf can use, although not with continuous efficiency, all wavelengths from about 4,000Å (an Ångstrom unit equals 1/10,000 of a micron) in the violet end of the spectrum to 8,000Å in the red. Plants do not effectively utilize the 10,000 to 12,000 foot-candles of light provided by bright sunlight, and photosynthesis may actualy be inhibited by over-illumination. When energy is being released by respiratory action as fast as it is being stored by photosynthesis, the *compensation point* is reached, and the plant neither gains nor loses. The compensation point for many plants is about 100 foot-candles, but will vary with changes in other factors, especially temperature. About 3% or less of the total energy of sunlight reaching a leaf is actually used in photosynthesis; the unused 97% is reflected, passes through the leaf, or is absorbed as heat.

THE CHLOROPLAST PIGMENTS AND ACCESSORY COMPOUNDS.

1. The chlorophylls, of which several are known, are the green chloroplast pigments. The most common of these substances is *chlorophyll a,* a bluish-green material found in all higher plants and in many algae. It is probably not a single compound. *Chloro-*

Fig. 10/4. Proposed structure of a molecule of chlorophyll a $(C_{55}H_{72}O_5N_4Mg)$. Four modified pyrrole rings are united into the larger porphyrin ring at the top of the figure. The phytyl portion of the molecule consists of the remaining carbon atoms strung out in a long chain, together with their associated hydrogen atoms.

phyll b, more yellowish-green than chlorophyll a, is present in lesser quantities. Some algae and photosynthetic bacteria have other chlorophylls: c, d, and e. All have a *porphyrin* ring, made up of smaller *pyrrole* rings, a magnesium atom, and a long chain of carbon atoms (Fig. 10/4), but the chlorophylls differ slightly from one another in their atomic arrangements.

2. Other pigments are closely associated with chlorophylls, but the way in which they function in photosynthesis has not been clearly established, although the fact of their collaboration has been ascertained. These pigments include the yellow or orange-yellow *carotenoids* (carotenes, xanthophylls) of higher plants, and the special pigments of algae: *phycoerythrin* of red algae and *phycocyanin* of blue-green algae.

3. Other chloroplast pigments, photosynthetically functional but produced in almost all living cells, are the various *cytochromes*. These iron-containing compounds act in the energy-transfer system by accepting and releasing electrons.

4. Also essential to the process are an array of specific enzymes, vitamins, and reducing agents. All are intimately bound, both in their physical structure and in their activities, by the lipoprotein framework of the grana, of their membranes, and of entire chloroplasts. The list of substances functioning in chloroplasts is still being expanded, and their exact roles in photosynthesis are being investigated.

ORGANIZATION OF THE PHOTOSYNTHETIC ORGANELLE. Photosynthesis apparently cannot occur if the necessary raw materials, light, and requisite biological compounds are merely brought together. An orderly structure is essential. In higher plants, this organized body is the chloroplast (Figs. 4/2, 10/5). In addition to its stacks of lipo-protein layers (the *grana*), chloroplasts have rounded, pebbly bumps, about 100Å in diameter which are visible only in high resolution electron micrographs. These are the *quantasomes*, and probably represent the smallest photosynthetic units. Still smaller, however, are the molecular arrangements of the chlorophylls, carotenoids, and other necessary compounds, but the way in which they are physically associated is not known.

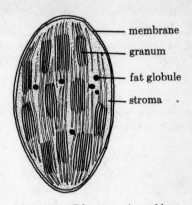

Fig. 10/5. Diagram of a chloroplast as seen in electron micrograph. (Magnification about 25,000.)

THE PROCESS OF PHOTOSYNTHESIS

Photosynthesis occurs in two recognizable but integrated steps. Indications of the existence of two steps come from such demonstrations as these: (1) Isolated chloroplasts may be illuminated in the presence of water and a hydrogen acceptor; e.g., a quinone, and oxygen will be liberated, but no sugar is formed. This is the *Hill reaction*. (2) Many cells, even animal cells in the dark, can fix carbon dioxide and synthesize carbohydrate if a source of chemical energy, adenosine triphosphate (ATP), is available.

Step One.   The first part of the total photosynthetic reaction involves the *absorption of light energy and the production of an energy-rich compound and a hydrogen carrier*. A chlorophyll molecule "excited" by the action of a photon of light transfers an energetic electron to an electron carrier, which in turn passes the electron on to still another acceptor. The result is the production of energy-rich adenosine triphosphate (ATP or A—P ～ P ～ P) by the addition of a third phosphate group to adenosine diphosphate (ADP, or A—P ～ P). (The symbol ～ indicates a high-energy chemical bond.) The formation of ATP by means of light energy is called *photophosphorylation.*

A second product of the "photo" portion of photosynthesis is the compound, reduced triphosphopyridine nucleotide, or $TPN \cdot H_2$. Acting as a hydrogen carrier, TPN is capable of receiving hydrogen from the water which was split during *photolysis,* thus becoming $TPN \cdot H_2$. This compound provides the hydrogen necessary for the synthesis of carbohydrate $(CH_2O)_n$ from CO_2 and H_2 in the dark phase (Step Two) of photosynthesis. Many biochemists refer to TPN as nicotinamide adenine dinucleotide (NAD).

Photophosphorylation is accomplished by two methods. In *cyclic photophosphorylation,* an excited electron from an illuminated chlorophyll molecule is passed along the electron acceptors, generating ATP, and is returned to the chlorophyll molecule from which it came. In *non-cyclic photophosphorylation,* two different chlorophyll molecules are activated. One furnishes electrons which are accepted by a series of carriers, with the resultant reduction of TPN by the addition of H_2 from photolysis; a second excited chlorophyll yields electrons which generate ATP, but which are then passed on to the first chlorophyll, thus restoring its lost electrons. The electrons lost from the second chlorophyll are restored from the water which was split. (Fig. 10/6)

After water is split, and the hydrogen from H_2O has passed to TPN to make $TPN \cdot H_2$, oxygen is released as molecular O_2. Experiments with oxygen isotopes proved that the oxygen liberated in photosynthesis is derived from water, and the oxygen in the new carbohydrate, say sugar $(C_6H_{12}O_6)$, comes from CO_2. The general formula for photosynthesis has long been considered, and for practical purposes still is

$$6\ H_2O + 6\ CO_2 \xrightarrow[\substack{chloro-\\phyll}]{light} C_6H_{12}O_6 + 6\ O_2$$

The fact that the liberated oxygen comes entirely from water makes a change necessary, so that a more nearly accurate statement of the over-all photosynthetic reaction is

$$12\ H_2O + 6\ CO_2 \xrightarrow[\substack{chloro-\\phyll}]{light} C_6H_{12}O_6 + 6\ O_2 + 6\ H_2O$$

Step Two. The second part of the total photosynthetic reaction involves *the incorporation of carbon dioxide into the final product.*

Like biological reactions in general, the reduction of carbon dioxide and the production of carbohydrate is a continuing process only so long as these conditions are met: (1) there must be a source of energy; (2) raw materials for the process must be brought to the site of reaction and the final products removed; (3) the reaction must be catalyzed by one or more enzymes; and (4) partially or entirely completed products, available from earlier times or even earlier cell generations, act as "primers."

In the reduction of CO_2 in a chloroplast, carbon is brought into the system via a series of reactions called the *Calvin cycle*. "Calvin" refers to one of the main discoverers of the scheme, and "cycle" indicates that an atom of carbon may pass through a number of reactions only to find itself, after a time, in a certain kind of molecule for a second or even a third time. Strictly speaking, however, metabolic cycles are impossible because no so-called cycle follows a predictable, rhythmically repetitious sequence, no atom regularly goes back to the same configuration that it formerly occupied, and time is always moving on. Nevertheless, for ease of understanding, many biological processes are called cycles and are diagrammatically so represented.

Using radioactive carbon as a tracer, and paper chromatography to isolate minute quantities of compounds, the path of carbon in CO_2 fixation has been followed. Cultures of algae are starved in the dark, then illuminated for short periods of time (fractions of a second to several seconds), and finally killed quickly. The cells are extracted and compounds are separated and identified.

The first compound to incorporate new CO_2 when photosynthesis starts is a three-carbon phosphorylated molecule: *phosphoglyceric acid*, $C_3H_5O_4 \cdot H_2PO_3$.

This compound is reduced to *phosphoglyceraldehyde*, $C_3H_5O_3 \cdot H_2PO_3$, by the action of $TPN \cdot H_2$ and ATP. Both these last-named compounds were produced during the water-splitting phase of photosynthesis. The oxidized TPN can be used again to carry another "load" of hydrogen. The ATP transfers energy during the process of splitting into ADP and inorganic phosphate; and the ADP is free to be phosphorylated again.

The phosphoglyceraldehyde, a 3-carbon compound, can be condensed to fructose, a 6-carbon sugar, and then to glucose, $C_6H_{12}O_6$, a sugar of such general biological usefulness, and so commonly made in photosynthesis that it is usually regarded as the end-product of the process even though no single compound can logically be so considered.

Not all the phosphoglyceraldehyde is converted to glucose. Some of it may react with the 6-carbon fructose to yield, not the expected 9-carbon result, but one *5-carbon* sugar (ribulose) and one *4-carbon* sugar (erythrose).

The 5-carbon ribulose, still carrying a phosphate group, can be further phosphorylated by ATP, thus becoming a 5-carbon sugar with two phosphate groups. This is *ribulose-diphosphate,* and is important because it is the compound which can pick up a new, incoming molecule of CO_2. The product of one molecule of ribulose-diphosphate plus one molecule of CO_2 is two molecules of phosphoglyceric acid.

Ribulose-diphosphate + carbon dioxide = phosphoglyceric acid

$$C_5H_8O_5(H_2PO_3)_2 \quad + \quad CO_2 \longrightarrow 2\,C_3H_5O_4 \cdot H_2PO_3$$

Thus, a sort of cycle is completed when more phosphoglyceric acid is made. The above "equation" is not balanced as far as hydrogen and oxygen are concerned because it is really a summary of several steps in which water is involved. The carbons do balance.

Meanwhile, a number of interconversions can occur, all leading eventually either to the production of more glucose (and from there to any of thousands of compounds) or to more ribulose-diphosphate. Some of these reactions are as follows:

(a)   3-carbon phosphoglyceraldehyde + 4-carbon erythrose =
7-carbon sugar (sedohepulose)

(b)   4-carbon erythrose + 6-carbon fructose =
two 5-carbon ribulose

(c)   7-carbon sedohepulose + 3-carbon glyceraldehyde =
two 5-carbon ribulose

The light reaction, including photophosphorylation and the production of $TPN \cdot H_2$ can be represented in a simplified scheme as shown in Fig. 10/6. The dark reaction involving the Calvin cycle can be represented, also in a simplified scheme, as shown in Fig. 10/7.

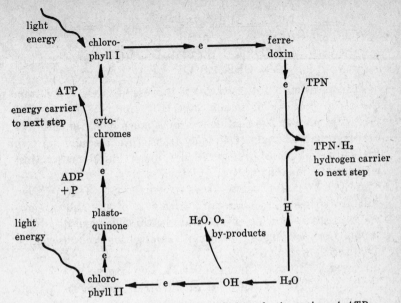

Fig. 10/6. *Non-cyclic photophosphorylation, the formation of ATP and the reduction of TPN following light-activation of two kinds of chlorophyll.*

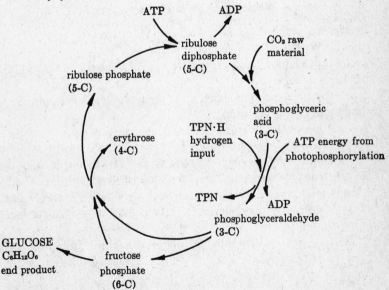

Fig. 10/7. *The uptake of carbon dioxide, the Calvin cycle, and some of the possible routes carbon may follow to the end product, glucose. The figures in parentheses indicate the number of carbon atoms in the named compounds.*

OTHER PHYSIOLOGICAL PROCESSES OF LEAVES

Transpiration. Transpiration is the loss of water vapor from the aerial portions of plants, especially from the leaves. The process is more or less continuous but is much more rapid during the day than at night. *Cuticular* transpiration occurs through the cuticle, *stomatal* transpiration through the stomata. Most water is lost in stomatal transpiration.

ADVANTAGES AND DISADVANTAGES TO THE PLANT. Transpiration pull is responsible in large part for ascent of sap and in some plants creates a cooling effect within leaves. The chief disadvantage of transpiration is that it can cause wilting and desiccation if it exceeds water absorption by the roots.

EXTERNAL FACTORS INFLUENCING TRANSPIRATION. (1) Atmospheric humidity, (2) light intensity, (3) air movement, (4) air temperature, 5) soil conditions.

Low humidity, high air temperature, bright illumination, and (within limits) moving air accelerate transpiration. Deficiency of water in the soil reduces the rate of transpiration.

INTERNAL FACTORS INFLUENCING TRANSPIRATION.

1. Heavy cutin layers reduce transpiration.

2. Perpendicular leaves transpire less than those in a horizontal position.

3. Stomata depressed below other epidermal cells transpire less than those exposed to wind.

4. Reduced leaf surface, and the shedding of leaves in dry seasons reduce transpirataion.

5. Opening and closing of stomata regulate transpiration to some extent but do not constitute a highly efficient control method. Closing occurs usually only after wilting has begun. Stomata control primarily oxygen and carbon dioxide exchange, not transpiration.

6. Presence of much colloidal material enables protoplasm to hold water tenaciously. Various gums and related substances are important in this respect.

The amount of water transpired by a plant, divided by its dry weight, is the *water requirement*. This varies in most plants from

200 to 1000. That is, for every pound of dry material manufactured by the plant, 200 to 1000 pounds of water are transpired.

Guttation. Guttation is the exudation of water in liquid form, usually through special pores called *hydathodes*. Guttation occurs usually from well-water plants on cool, moist nights following warm days. Guttation is probably of little physiological significance.

Leaf-fall and Autumnal Coloration. In autumn, a special layer of cells, the abscission layer, may form at base of petiole. Cells behind this layer become corky and impervious to water. The normal green color of a leaf is due to the pigment *chlorophyll*. Since chlorophyll is decomposed by light it disappears when conditions for its renewal are unfavorable, allowing other already-present pigments (yellow *xanthophyll* and orange *carotene*) to become evident. Deep red colors are new pigments (anthocyanins) which develop in unknown ways as chlorophyll disappears; deep red colors are associated with high sugar content, low temperature, and high light intensity. The middle lamellas of abscission cells disintegrate, the cells separate, and the petiole breaks off where this happens. A cork layer below the abscission layer prevents water loss from the leaf scar.

SPECIALIZED (MODIFIED) LEAVES

In some plants, some or all of the leaves have become modified structurally, so that they scarcely or in no way resemble typical foilage leaves. These structural modifications are correlated with certain specialized functions which these modified leaves perform. Common types of specialized leaves are:

Bud Scales. These are overlapping, modified leaves which protect the internal growing tissues of the buds of most plants growing in regions of low winter temperatures or of great aridity.

Spines. Spines with buds in their axils are modified leaves. They may protect the plant from grazing animals.

Bulb Scales. In such underground stems as the tulip and onion, bulb scales serve for food storage.

Water Storage Leaves. In succulent plants (stonecrop, live-forever), these fleshy structures with thick layers of cutin serve for the storage of water.

Tendrils.   Some plants (sweet pea, garden pea) use these whole leaves or parts of leaves for climbing.

Insect Trapping Leaves.   The pitcher plants, sundew, and Venus' fly-trap have these highly specialized leaves which attract, capture, and digest insects which supplement their normal nutritional requirements.

Reproductive Leaves.   Leaves that perform reproductive functions occur in many groups of plants. Many botanists interpret some of the parts of flowers, pine cones, etc., as leaves which function primarily for reproduction. Foliage leaves of some plants (*Bryophyllum*) are able to produce new plants from their petioles or marginal notches.

Chapter 11

Metabolism: Digestion and Respiration

The term *metabolism* includes all the chemical transformations that take place in cells or organisms. Metabolism entails food manufacture, food transformations, release of the energy stored in foods, construction of protoplasm and of cell walls, and reproduction. Numerous and complex processes are involved in metabolism, and detailed knowledge of them presupposes extensive knowledge of chemistry and physics. We still know very little about basic factors of most biological activities.

FOODS

Photosynthesis is the fundamental process of food manufacture. From the sugar made in photosynthesis and from mineral salts, all other kinds of foods are made.

A food is an organic material from which living things derive energy or build protoplasm. All foods contain carbon, hydrogen, and oxygen, which, through photosynthesis, are built into sugars from carbon dioxide and water. Some foods contain additional elements obtained from mineral salts (nitrates, phosphates, sulfates) absorbed by roots from soils. Plants also require the elements listed in Chapter 5, but not all the elements are built into foods as defined above. Special compounds, called vitamins or hormones, may perform regulatory functions. Certain elements are absorbed by plants in considerable quantities and must be restored to soils through fertilizers made chiefly from bone meal, blood, manure, dead leaves, peat, guano of birds, special rocks, or synthetic chemicals. Plants can be grown in jars containing water solutions of mineral elements, or in gravel or sand containing such solutions. Growing plants in such cultures is *hydroponics*.

Animals depend on plants for food. Animals are able to synthesize fats and proteins from sugars and nitrogenous compounds, but they are unable to make foods from inorganic materials as plants can.

There are three groups of foods: (1) *carbohydrates*, (2) *fats and oils*, (3) *proteins*. Their characteristics and functions:

Carbohydrates.   These foods contain carbon, hydrogen, and oxygen, with the hydrogen and oxygen in the same proportion as they are in water, namely two to one. Some carbohydrates are water-soluble (sugars), others are not (starch and cellulose). The functions of carbohydrates in plants are: to supply energy (sugars), to build cell walls (cellulose), and to build other kinds of foods. Carbohydrates are important energy foods and are stored in many places in plants, chiefly as starches. These are abundant in roots, tubers, fruits, etc. Common carbohydrates are *glucose* ($C_6H_{12}O_6$), *sucrose*, or cane sugar ($C_{12}H_{22}O_{11}$), *starch* $[(C_6H_{10}O_5)n]$, and *cellulose* $[(C_6H_{10}O_5)n]$.

Fats and Oils.   These contain carbon, hydrogen, and oxygen, with proportionately less oxygen as compared with the oxygen content of carbohydrates. Carbohydrates can be enzymatically converted to fats in the following manner. A molecule of glucose can be made to yield *glycerol*, $C_3H_8O_3$, an alcohol which is an essential part of many fats. By a different metabolic pathway, sugar can be made to yield a variety of *fatty acids*, frequently having the general formula $C_nH_{2n+1}COOH$. The condensation of one molecule of glycerol with three molecules of fatty acids yields a fat molecule called a *triglyceride*. Many fats and oils are triglycerides, but others may be made with different alcohols than glycerol, or they may be bonded to phosphoric acid groups in a great variety of compounds. Fats and oils are similar chemically, but fats are solids at room temperatures, while oils are liquids. All fats and oils are greasy and are water-insoluble. They occur in all living cells, but are especially abundant in storage organs such as seeds (peanut, soybean) and fruits (banana, avocado). Fats are reserve foods which can be called upon to supply energy, and they form an integral part of the extensive membrane systems that pervade living cells. One of the main reasons that fat solvents are toxic is that they destroy essential membranes and consequently disrupt normal metabolic activities.

Proteins. Proteins have the largest and most complex molecules in the biological world. They have molecular weights in the millions, occur in countless variations, and, with the exception of nucleic acids, proteins are the only compounds present in all biological entities, including even viruses. *Simple proteins* can be broken down (digested, hydrolyzed) to *amino acids.* Indeed, a protein is a collection of amino acids that are connected in a special way.

An *amino acid* is one of a number of organic compounds which have: (1) a body of atoms which may be a simple methyl group, CH_3, or a complex of rings of atoms; (2) a carboxyl group, $-COOH$, which gives the molecule its acid character; and (3) an amino group, $-NH_2$, which gives the molecule a basic character. The amino acid molecules and the proteins made from them thus have capabilities of acting either as acids or bases, depending upon the acidity or alkalinity of the surrounding medium. The biologically important amino acids, of which about two dozen are common, all have the general configuration

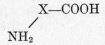

The X indicates a group of atoms to which the amino and carboxyl groups are bonded. The amino group is attached to the carbon atom nearest the carboxyl group. That carbon atom is called the *alpha carbon,* and such amino acids are *alpha amino acids.*

Two or more amino acids can be connected by *peptide bonds* if a hydroxyl group ($-OH$) of one acid, and a hydrogen of a second acid are removed. The two molecules are connected as shown in Fig. 11/1.

Two amino acids thus joined form a *dipeptide;* if a few acids are involved a *polypeptide* is formed; and if a large number combine a simple protein is formed.

Under the action of strong chemicals or of enzymes, the peptide bond can be broken, the H_2O that was lost in the formation of the peptide bond is replaced (hydrolysis), and the original amino acids are released.

The individuality of the countless proteins in the world is due to the arrangement of the amino acids and to the twists and cross-

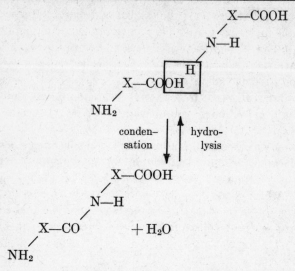

Fig. 11/1.   Two amino acid molecules.

linkages, or lack of them, in the amino acid chain. Usually twenty amino acids are available for construction, and a protein molecule can contain hundreds of them in any combination. The number of possible kinds of protein molecules is consequently large enough for each individual on earth to have its own proteins with little chance of their duplication in another individual.

Additional complexities in protein structure result from the fact that some amino acids contain sulfur, which is capable of forming cross-bonds between amino acids in addition to the peptide bonds. This kind of bonding tends to keep the protein molecules from being straight chains, but makes them double back, resulting in special molecular shapes. In addition, proteins may be *conjugated* with molecules other than amino acids: pigments, lipids, phosphates.

In plants, proteins are synthesized in higher concentrations in reproductive parts than in vegetative ones, and are especially abundant in seeds.

The biosynthesis of protein will be treated as a special topic, as an aspect of growth in Chapter **13**.

Methods of Getting Food. Autotrophic plants are those which manufacture their own food from inorganic materials.

There are two kinds of autotrophic plants: *photosynthetic,* those which contain chlorophyll and which make food using the energy of light; and *chemosynthetic,* which lack chlorophyll and which obtain their energy for food synthesis by the oxidation of hydrogen sulfide, ammonium compounds, etc. There are several kinds of chemosynthetic bacteria. *Heterotrophic plants* are those which are unable to manufacture their own food and which depend upon previously synthesized foods for their energy and their protoplasm-building. There are two main types of heterotrophic plants: *parasites,* which take their food directly from other living organisms; and *saprophytes,* which get their food from the dead remains or waste products of organisms. Some organisms can live both saprophytically and parasitically. Others (e.g., mistletoe) have some chlorophyll and are thus partly autotrophic, and partly heterotrophic. An example of a parasite is the wheat-rust fungus; of a saprophyte, a mushroom living on dead leaves and other decomposing organic matter in soils.

CATABOLIC PROCESSES

The metabolic processes thus far described are processes of conversion of simple substances into more complex ones. Proceeding simultaneously with these constructive processes are the *catabolic* processes of *digestion* and *respiration* which involve transforming complex organic compounds to simpler substances.

Digestion.   Digestion is the process in which water-insoluble, non-diffusible foods are converted into water-soluble, diffusible foods, or in which complex foods are converted into simpler ones. Digestion is usually necessary before translocation, respiration, and certain other processes can occur. In most higher plants, digestion takes place inside cells, but in bacteria, fungi, and some parasites, digestion may be extracellular, with the soluble endproducts diffusing into or being actively brought into the enzyme-producing cells. Digestive processes, in general, are hydrolytic.

Digestion involves the uptake of water in transforming complex into simpler foods, as shown above in the example of proteins and amino acids, or below in the digestion of maltose:

$$\underset{\text{(maltose)}}{C_{12}H_{22}O_{11}} + \underset{\text{(water)}}{H_2O} \xrightarrow{\text{maltase}} \underset{\text{(glucose)}}{2\ C_6H_{12}O_6}$$

Digestion does not occur spontaneously but requires the action of a digestive agent, or *catalyst,* which initiates and controls the process, or at least speeds it enormously. A molecule of cellulose might eventually break down to its glucose components (*residues*) without outside help, but the chances of that happening are ridiculously small. The organic catalysts which are produced by protoplasm and which control digestion are termed *enzymes.* Enzymes are named from the *substrate* upon which they are capable of working, plus the suffix *-ase.* The characteristics of enzymes are:

1. They are not used up or changed in the processes in which they are involved.

2. They are efficient. A small quantity of enzyme can perform a large quantity of conversion.

3. They are destroyed (*denatured*) by heat at or near the boiling point of water, and are temporarily inactivated by low temperature, near the freezing point of water.

4. They are complex, colloidal materials, mostly proteinaceous, and frequently have non-protein *coenzymes* as auxiliary agents.

5. They are specific, each enzyme acting upon a single substrate or on a few atomically similar substrates.

6. Their action is in most cases reversible; e.g., they can build up complex foods from simple foods, and vice versa.

7. Their action is affected by their environment; e.g., the acidity of the solution, the concentration of the substrate, and the concentration of the end products.

Some of the digestive (hydrolytic) enzymes in plants are:

1. *Diastase,* a mixture of enzymes which convert starch into maltose.

2. *Lipases,* which hydrolyze fats into fatty acids and glycerol.

3. *Sucrase,* which hydrolyzes cane sugar into glucose and fructose.

4. *Proteases,* which hydrolyze proteins into polypeptides and amino acids.

5. *Cellulase,* which hydrolyzes cellulose to cellobiose.

Respiration.   Respiration is a general term covering a varied, complex series of biological phenomena by means of which the chemical energy of foods is transferred to the chemical energy of some compound, usually adenosine triphosphate (ATP). Cells use ATP to do work; therefore, this compound is referred to as "the energy currency of cells."

The over-all equation for respiration is

$$C_6H_{12}O_6 + 6\ O_2 \longrightarrow 6\ CO_2 + 6\ H_2O + energy$$

The energy may be used in chemical syntheses, movement, various growth processes, or it may be liberated as heat, light, or electrical energy. Although the energy contained in food may be made available for general work in a variety of ways, one common method is via *glycolysis*, the *Krebs cycle*, and *terminal oxidations*.

GLYCOLYSIS.   Glucose is a compound which almost all cells can use both as a starting material for building other compounds and for the production of ATP. The initial steps in the utilization of glucose for energy transfer occur as a result of enzyme action in cytoplasm. The first step is the phosphorylation of the glucose molecule, an action by means of which the glucose receives a phosphate group from a molecule of ATP which thereupon becomes ADP. An enzymatically mediated rearrangement of the glucose phosphate results in fructose phosphate, which is further phosphorylated to fructose diphosphate at the cost of another ATP. This 6-carbon molecule is split into two 3-carbon molecules of phospho-glyceraldehyde, each of which is further phosphorylated by inorganic phosphate, and oxidized by transferring a hydrogen atom to a hydrogen carrier. The hydrogen carrier is diphospho-pyridine nucleotide, DPN (synonymous with nicotine adenine dinucleotide, or NAD), which is reduced by the hydrogen to DPNH. By this time, the cell has "spent" two molecules of ATP, and has on hand one DPNH molecule and two molecules of diphosphoglyceric acid. One phosphate group is used to generate a molecule of ATP, and the cell has thus gained back its two invested ATP's and still has two molecules of phosphoglyceric acid. The compounds up to this point are the same as those involved in the CO_2 cycle in photosynthesis. The two phosphoglyceric acid molecules next generate another molecule of ATP each, making a total cost of two ATP's and a formation of four. The cell's total gain, then, is two ATP's with a resultant pair of pyruvic acid molecules, and the DPNH still in its reduced form.

These events, with much detail omitted, take place without any uptake of oxygen, or any liberation of CO_2 or water. Most organisms follow a somewhat similar scheme up to this point, the entire process being called glycolysis. Glycolysis can be diagrammatically shown as follows:

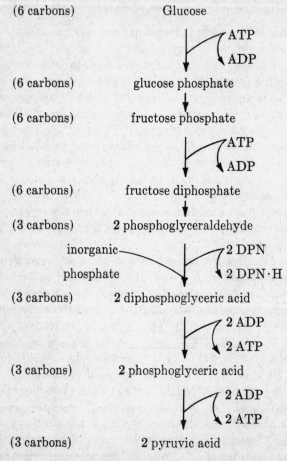

(6 carbons)   Glucose

ATP / ADP

(6 carbons)   glucose phosphate

(6 carbons)   fructose phosphate

ATP / ADP

(6 carbons)   fructose diphosphate

(3 carbons)   2 phosphoglyceraldehyde

inorganic phosphate   2 DPN / 2 DPN·H

(3 carbons)   2 diphosphoglyceric acid

2 ADP / 2 ATP

(3 carbons)   2 phosphoglyceric acid

2 ADP / 2 ATP

(3 carbons)   2 pyruvic acid

Fermentation. Some organisms characteristically cease obtaining chemical energy at the end of the glycolytic sequence, and use the DPNH to reduce pyruvic acid to other organic compounds, with the release of CO_2. Some compounds thus formed by respiration without oxygen uptake are acetaldehyde, ethyl alcohol, acetic acid, and a variety of other alcohols and acids. The bubbles of CO_2 in bread, beer, and sparkling wines are the result of such *fermentation*, in which no atmospheric oxygen is used, and only a small amount of the energy available in the original sugar is transferred to the chemical bond energy of ATP. Most higher plants are capable of living temporarily on fermentative production of ATP; but yeasts and bacteria use the method more than green plants do. Fermentation is an inefficient means of energy transfer, allowing the cell to take advantage of less than 5% of the energy available in the food. The rest is lost as heat, or remains in the form of chemical bonds in the incompletely oxidized organic compounds.

Anaerobic respiration.   Some bacteria respire in the absence of atmospheric oxygen, utilizing some inorganic material, for example iron or sulfur, as hydrogen acceptors. Like fermentation, strict anaerobic respiration results in low ATP formation and in incompletely oxidized foods. Fermentation and anaerobic respiration are similar except in the matter of whether the hydrogen acceptor is organic (pyruvic acid in yeast) or inorganic (sulfate in some bacteria), and both terms can be treated as synonyms.

THE KREBS CYCLE.   Also known as the citric acid cycle or the carboxylic acid cycle, this is a series of chemical changes in the course of which pyruvic acid is broken down to CO_2 and H, with subsequent oxygen uptake and energy transfer to ATP. The Krebs cycle is essentially a means of feeding two carbon atoms at a time into an enzyme system, with energy which was earlier bound in photosynthesis transferred to the bond energy of ATP. By far the greater part of the energy of foods is made available by this system.

The compound which feeds energy-containing "fuel" into the Krebs cycle is pyruvic acid. It is combined with a molecule of the sulfur-containing, vitamin-like Coenzyme A (CoA) and loses a molecule of CO_2. The result is a new molecule, acetylCoA, which contains the CoA and two carbon atoms (an acetyl group) from the pyruvic acid. AcetylCoA is a key compound in respiration because through it most foods, whether they are carbohydrates, fats, or proteins, may be funneled into the Krebs cycle.

The two acetyl carbon atoms of acetylCoA are next united with a four-carbon acid (oxalacetic acid) present in the cell, to form a molecule of six-carbon citric acid, and CoA is released for another carbon-carrying trip.

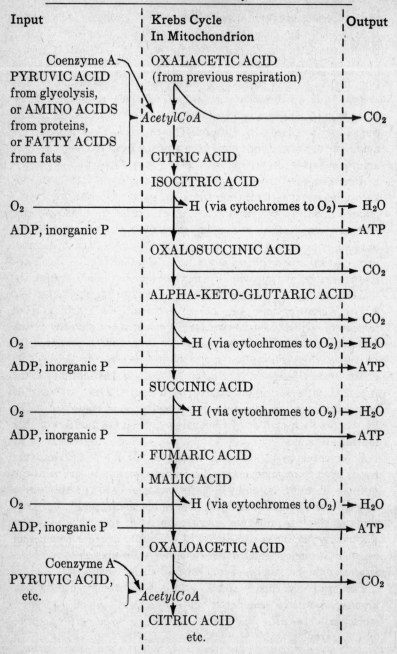

Input	Krebs Cycle In Mitochondrion	Output
Coenzyme A	OXALACETIC ACID (from previous respiration)	
PYRUVIC ACID from glycolysis, or AMINO ACIDS from proteins, or FATTY ACIDS from fats	*AcetylCoA*	→ CO_2
	CITRIC ACID	
	ISOCITRIC ACID	
O_2	H (via cytochromes to O_2)	→ H_2O
ADP, inorganic P		→ ATP
	OXALOSUCCINIC ACID	
		→ CO_2
	ALPHA-KETO-GLUTARIC ACID	
		→ CO_2
O_2	H (via cytochromes to O_2)	→ H_2O
ADP, inorganic P		→ ATP
	SUCCINIC ACID	
O_2	H (via cytochromes to O_2)	→ H_2O
ADP, inorganic P		→ ATP
	FUMARIC ACID	
	MALIC ACID	
O_2	H (via cytochromes to O_2)	→ H_2O
ADP, inorganic P		→ ATP
	OXALOACETIC ACID	
Coenzyme A		→ CO_2
PYRUVIC ACID, etc.	*AcetylCoA*	
	CITRIC ACID etc.	

Fig. 11/2. Simplified scheme of Krebs cycle in mitochondrion, with input of energy-containing organic compounds, release of carbon dioxide, uptake of oxygen, and synthesis of adenosine triphosphate. Each passage of H along the cytochrome system generates three molecules of ATP.

A series of enzymes then generally brings about the following sequence. (1) Citric acid is rearranged internally to form isocitric acid. (2) Isocitric acid loses hydrogen, becoming oxalosuccinic acid. TPN accepts the hydrogen. (3) Oxalosuccinic acid loses a molecule of CO_2 to become alpha-keto-glutaric acid. (4) Alpha-keto-glutaric acid loses CO_2 and H to become succinic acid. (5) Succinic acid loses H to become fumaric acid. (6) Fumaric acid plus water forms malic acid. (7) Malic acid loses H to become oxalacetic acid. (8) Oxalacetic acid is the compound which can condense with two carbon atoms from acetylCoA, as occurred at the beginning of the cycle.

THE TERMINAL OXIDATIONS. The CO_2 given off during the Krebs cycle is released as a gas and diffuses out of the cell. The hydrogen atoms which were given off during the cycle are transferred along a chain of hydrogen acceptors, having been removed by a series of oxidizing enzymes, the *dehydrogenases*. The hydrogen is passed along via DPN or TPN to the *cytochromes*, the usual sequence being cytochrome b, c_1, c, a, and a_3; the chain of reactants is called the *cytochrome system*. Hydrogen is finally united with oxygen to produce the "water of respiration." The material end products of respiration are thus CO_2 and H_2O. In their passage along the hydrogen carriers, the hydrogen atoms (or more exactly, electrons) transfer their energy in an unknown manner to the chemical bonds of pyrophosphate. These bonds are the ones formed when ADP plus inorganic phosphate form ATP.

This method of producing ATP is known as *oxidative phosphorylation*, in contrast to the *photophosphorylation* that occurs in chloroplasts. Oxidative phosphorylation is the main energy-transferring mechanism in most cells, and, except for some microorganisms which lack them, the process occurs in the *mitochondria*.

Photosynthesis and Respiration Contrasted. A recapitulation of the general features of photosynthesis and respiration shows the following contrasts:

Photosynthesis	*Respiration*
1. Takes in carbon dioxide	1. Releases carbon dioxide
2. Releases oxygen	2. Binds oxygen
3. Synthesizes sugars and other organic compounds	3. Degrades sugars and other organic compounds
4. Results in increase in dry weight	4. Results in decrease in dry weight
5. Occurs only in the presence of chlorophyll	5. Occurs in all living cells
6. Stores energy	6. Releases energy
7. Occurs only when light energy is available	7. Occurs throughout entire life of any cell

Chapter 12

Growth: Cell Division
and Differentiation

Growth is a result of the many metabolic processes of plants. It involves (1) the formation of new cells, (2) the quantitative increase of these cells, and (3) the maturation or differentiation of cells. Growth thus involves the manufacture of food and other substances, digestion, respiration, etc. Growth usually results in an irreversible gain in size and weight and a more or less irreversible differentiation of cells, tissues, and organs. If a curve is plotted to indicate the growth in size of an organism at regular intervals, the curve is S-shaped, indicating that growth begins slowly, then passes into a phase of rapid enlargement, following which the growth rate gradually decreases until the point of cessation. If, however, rate of increase instead of total increase is plotted against time, a bell-shaped, or *cloche*, curve results. Similar curves are obtained from growing individuals or from populations of individuals, as when a species is introduced into a new environment or when a microorganism is used as a start of a laboratory culture. The slow start of growth is the *lag phase*, and the ensuing phase of rapid growth is the "grand period of growth" or, in instances in which growth is logarithmic, the *log phase*.

Growth and reproduction are inseparable. Growth of an individual higher plant involves cell division, which is a reproduction of cells. If some cells are removed from the individual and continue to reproduce themselves, a new individual results. The cells separated from a parent plant may or may not be associated with cells from a second parent; that is, reproduction may be sexual or non-sexual. The basic phenomena of reproduction remain, essentially, the growth of cells and the increase of cell numbers, regardless of what special features may accompany the action.

In single-celled organisms, cell division and reproduction are identical.

Growth in higher plants occurs chiefly in buds, root tips, cambium, cork cambium, and root pericycle. In all these regions, the same series of events occurs: cell division, enlargement, and differentiation.

CELL DIVISION

Except for bacteria and blue-green algae which lack membrane-bound nuclei, plant cells increase in number by the complex method of *mitosis*. The morphological aspects of the process have been known since the 1860's. Some of its submicroscopic and molecular features have been elucidated during the recent past, and knowledge of mitosis is increasing rapidly, but much remains to be learned. Some important unanswered questions are concerned with the forces that initiate mitosis, fix the direction of division, drive the chromosomes through the cytoplasm, determine how and where and when new walls are deposited, and whether a division will be a regular mitotic one or a reductional one.

For convenience, biologists speak of mitosis as being divided into four phases, but the process is a smoothly continuous one with an infinite number of possible stages, with some of the important ones not discernible by direct observation (Fig. 12/1).

Interphase. This phase is also called the metabolic phase, and is sometimes erroneously called the "resting phase." During interphase a cell is doing its general work for the organism. The activity which is important for cell division is the increase in nuclear material. The deoxyribose nucleic acid (DNA) content of a nucleus is doubled in quantity during interphase. This is therefore the time when the chromosomes are being replicated, even though this action cannot be seen in the microscope in either living or fixed material. In interphase, nucleoli are present, and the nucleus is surrounded by a two-layered membrane containing pores (Fig. 4/2).

Prophase. At the onset of mitosis, the nuclear membrane is intact and the chromosomes are so long and thin as to be invisible by ordinary optical means. During later prophase, the nuclear membrane is removed into the cytoplasm as part of the

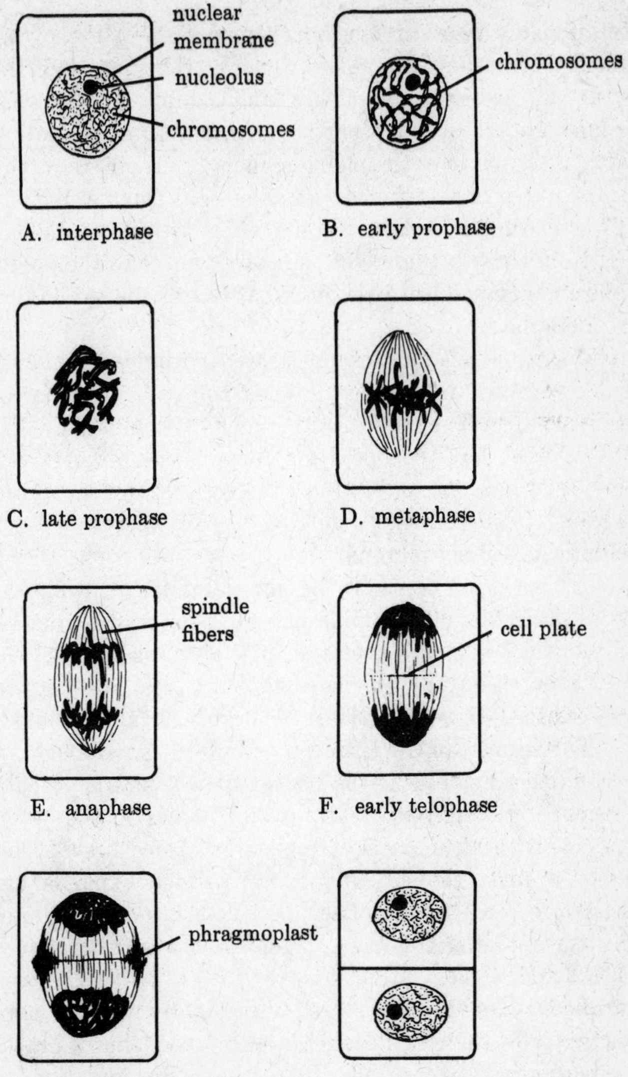

A. interphase

nuclear
membrane
nucleolus
chromosomes

B. early prophase

chromosomes

C. late prophase

D. metaphase

E. anaphase

spindle
fibers

F. early telophase

cell plate

G. late telophase

phragmoplast

H. two cells

Fig. 12/1. A mitotic division of a plant cell.

endoplasmic reticulum. The chromosomes shorten and thicken, and the nucleolus or nucleoli break up.

Metaphase. The chromosomes, lying free in the cytoplasm, are lined up in a flattened disk or ring at the *equator* of the cell. They are all present in duplicate, and exhibit any individuality they may have, such as constrictions, satellites, variability of staining (*heterochromatin*), long or short arms, etc. Each chromosome, except in rare instances, has a special spindle fiber attachment region, the *centromere*. Spindle fibers are microtubules which extend from the centromere of the chromosome to the pole of the cell and in aggregate form a somewhat barrel-shaped body in the cell at metaphase.

Anaphase. Each pair of duplicated chromosomes separates, with one member moving toward one pole, and its twin moving toward the other. The movement is quick, as compared with other mitotic phases, but the operating force is unknown. While the chromosomes are en route from the equator to the poles, the poles themselves are usually moving further apart.

Telophase. When chromosomes arrive at a pole, they come close together and cease to be individually identifiable. The endoplasmic reticulum rebuilds nuclear membranes, and certain chromosomes, perhaps all, engage in reforming nucleoli. At the equator a partition, the *cell plate*, is being laid down, separating the cytoplasm into two portions. A microscopically visible structure, which seems to help form the cell plate, moves outward like an expanding ring. This is the *phragmoplast*, and is probably an expression of the activity of the *Golgi bodies* in the cytoplasm (Fig. 4/2). A sheet of watery droplets is formed first; then these coalesce to form the cell plate itself which expands until it reaches from wall to wall. Then new depositions of cellulose on each surface of the cell plate complete the new primary wall (Figs. 4/2 and 12/1).

Significance of Mitosis. Cell division by mitosis not only makes two cells where earlier there was only one, but it practically guarantees genetic constancy from one cell generation to the next. When chromosomes duplicate themselves during interphase, they do it accurately, so that the two chromosomes of the new pair are chemically and geometrically identical. Then, when anaphase movement begins, each chromosome pair separates. The result is that at telophase, the two new nuclei have precisely identical chromosome sets, or *genomes*. As long as mitotic divisions occur normally, the cell population of an organism is genetically uniform. (Abnormal mitoses are, however, commoner than was once thought, and vegetative cells with aberrant genomes are not uncommon.)

Cytokinesis. The cell plate formed at telophase divides the cytoplasm of the old cell into usually about equal parts. The separation of cytoplasm (*cytokinesis*) may or may not directly follow nuclear division (*karyokinesis*). In some organisms (some algae and fungi), cytokinesis occurs only in reproductive parts, and even in flowering plants, some parts (the "milk vessels" or latex system) may not lay down cell walls between nuclei.

When cytokinesis does occur, the cytoplasmic particles are more or less equally divided between the two parts. Some of the particles, especially plastids and mitochondria, are themselves capable of dividing and thus keep up their numbers during repeated cell divisions. These particles have their own reproductive systems and their own DNA, and are only in an indirect way under the control of the nuclear DNA.

MEIOSIS

Meiosis is a special kind of mitosis, which it resembles in some ways, namely in the formation of spindle fibers, in the general aspect of the division figures, and in the multiplication of cells. But as far as the chromosomes are concerned, meiosis differs from mitosis, and an understanding of meiotic events is essential to an understanding of growth, sexuality, genetics, and evolution.

In all plants except the Thallophytes, meiosis takes place just before the production of spores. In flowering plants, meiosis in the anther leads to the formation of *microspores*, and in the ovary it leads to the formation of *megaspores*. Both kinds of spores will in a few cell generations produce *gametes* or sexually functional nuclei.

A vegetative nucleus, with its full complement of chromosomes, contains one set of chromosomes which descended through various cell generations from the female parent of the organism, and a second set from the male parent. These two sets of chromosomes are generally similar in number, size, shape, structural detail, and

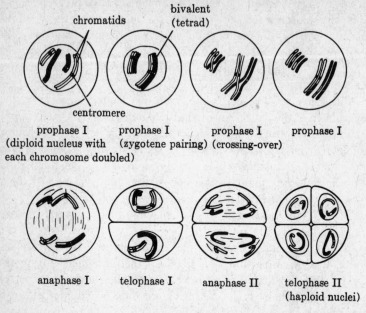

Fig. 12/2. Meiotic divisions. Selected stages show crossing-over and reduction of chromosome number from four to two.

especially in their genetic or biochemical arrangements. In Fig. 12/2, the dark chromosomes represent the maternal and the light ones the paternal set. In this digrammatically simplified instance there are two pairs from each parent: a pair of long ones with a central centromere, and a pair of short ones with an almost terminal centromere. Each chromosome member of a pair is referred to as a *homologue*.

Meiosis requires two divisions, each superficially resembling mitosis in its observable phases.

In the prophase of the first division (Prophase I), each member of a pair of homologues is duplicated and the two members pair up with such exactitude that each part of a chromosome is matched precisely with the corresponding part of its homologue. This is the *zygotene* stage, sometimes called *synapsis*. Here the "chromosome" is four-stranded, with two strands in each of the matched homologues, and is called a *bivalent* or a *tetrad*. Each strand is a "half-chromosome" or *chromatid*, and identical chromatids are *sisters*. One important difference between mitosis and meiosis is that in mitosis the homologues are not associated in pairs as they are in meiosis.

Also during Prophase I, non-sister chromatids exchange pieces by a system of crossing-over. In Fig. 12/2, for the sake of simplicity, only one crossover is shown in each bivalent, although this is not typical, and of course such crossovers are not actually so obvious as they are in diagrams.

Metaphase I, Anaphase I, and Telophase I follow, with two nuclei resulting. Each new nucleus apparently has only half as many chromosomes as the microspore (or megaspore) mother cell had, but each chromosome is double. The nuclei immediately pass into Prophase II, followed by the other usual phases, but without any further chromosome duplication. When Telophase II is over, four nuclei are present, each with one set of chromosomes, even though (as a result of crossovers) the strands are not exact copies of the original chromosomes. This is in contrast to the chromosome situation in the vegetative nuclei which contained two sets of chromosomes. A nucleus with two sets, or two genomes, is said to have a double, or diploid, or 2n, or 2x chromosome complement, as opposed to the single, or haploid, or n, or x number.

The important features of meiosis are (1) the reduction of the chromosome number from diploid to haploid, and (2) the reshuffling of genetic material during crossing-over. In flowering plants, the product of the reduction divisions in the anther is a quartet (*tetrad*) of microspores; in the ovule, four megaspores. From a microspore will come pollen and eventually male gametes, or sperms; from a megaspore will come an embryo sac and eventually a female gamete, or egg.

CELL ENLARGEMENT

After a mitotic division, the combined volume of the two new cells is about equal to the volume of the parent cell. If general growth in size is to take place, each cell must increase in volume. The bulk of the increase comes in the vacuoles, which are small and numerous in young cells, but larger and fewer in older ones. Many cells increase as much as 500 times in volume as they age, with most of the increase being due to water intake in the vacuoles. Meanwhile, although it increases relatively less, the cytoplasm is increasing, too. At full size, the cytoplasm may form no more than a thin coating between the vacuole and the cell wall. Of course the wall must stretch extensively, and new wall material in the form of interwoven cellulose microfibrils is incorporated. The wall formed by a cell during its enlargement phase is the *primary wall* (Figs. 4/1 and 4/2). The final shape of the cell is determined in part by the original orientation of its microfibrils, but detailed explanations for the ultimate size and shape of cells are lacking or at best sketchy.

CELL DIFFERENTIATION

After a cell has grown to full size, it will in many tissues undergo a final specialization. The most obvious changes are in the thickening, sculpturing, or special chemical composition of the walls. The last wall layers to be deposited when the stretching of the primary wall has ceased are called *secondary walls*. The walls have several layers, and may be impregnated with a variety of substances besides cellulose; they may be pitted or perforated, or may be fitted with various kinds of structural reinforcement (Chap. 4). Less apparent, but equally important, are cytoplasmic specializations which differ according to the position of the cell in the plant, for cells may be secretory, photosynthetic, storage, reproductive, or absorptive cells, and each has its characteristic organization. We do not know how the terminal differentiations are determined or brought about.

ALTERNATION OF GENERATIONS

During the development of plants through the centuries, from ancient, simple types to more recent, complex ones, a system of alternating sexual and non-sexual phases has evolved. The phase which produces gametes, or sex cells, is the *haploid phase* and is called the *gametophyte generation*. When two gametes unite, these haploid cells in fusion form a so-called "fertilized egg," or zygote, which is, as a result of the meeting of two haploid nuclei, diploid. A zygote is the first cell of a new *diploid phase* which will end with the formation of spores, and is called the *sporophyte generation*. Reduction division occurs immediately before spore formation, so that spores are haploid, and represent the first cells of a new gametophyte generation. Thus the chromosome complement of a plant can change at two periods: (1) when two haploid gametes meet in fertilization to initiate the diploid sporophyte, and (2) when meiosis results in the production of haploid spores to initiate the haploid gametophyte.

In its simplest form, the alternation of generations may be represented by the following diagram:

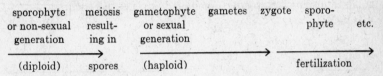

sporophyte   meiosis   gametophyte   gametes   zygote   sporo-
or non-sexual   result-   or sexual   phyte   etc.
   generation   ing in   generation

(diploid)   spores   (haploid)   fertilization

In some of the simplest plants, as algae, the haploid gametophyte generation is the main phase in size and duration, with the diploid sporophyte generation lasting for only the life of one cell before reduction division restores the haploid phase. Such an alternation of generations could be represented thus:

sporophyte   gametophyte   sporophyte   etc.

In mosses, the gametophyte generation is the familiar moss plant. It is somewhat larger and lives longer than the sporophyte, and is nutritionally independent. Its "life cycle" could be represented thus:

partially   independent   partially
dependent   gametophyte   dependent

sporophyte   the moss   sporophyte   etc.
                        plant

(See Chap. 21, Fig. 21/1.)

In ferns, the familiar fern plant is the sporophyte. It is much larger and longer-lived than the gametophyte. Its "life cycle" could be represented thus:

sporophyte   gametophyte   sporophyte

the fern plant   (prothallus)   the fern plant   etc.

(See Chap. 23, Fig. 23/1.)

An additional complication was introduced into the scheme with the evolution of two different kinds of spores, large *megaspores*, which give rise to female gametophytes capable of producing eggs only, and small *microspores,* which give rise to male gametophytes capable of producing sperms only. Life cycles of plants producing both megaspores and microspores, a condition known as *hetero-spory* as contrasted with the *homospory* of most ferns, are typical of *Selaginella* (Chap. 22, Fig. 22/1), and of all seed plants (Gymnosperms and Angiosperms, Chaps. 15, 16, and 23). A cycle involving heterospory could be represented thus:

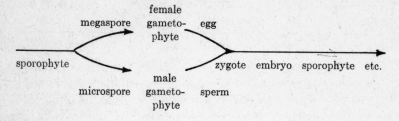

Chapter 13

Growth: Molecular Action

Deoxyribose nucleic acid, DNA, has been singled out as *the* genetically critical material because: (1) genetic and cytological experiments showed that chromosomes are the carriers of genetic control from cell to cell and from generation to generation; (2) DNA is limited almost entirely to chromosomes; (3) DNA is the only component of chromosomes which remains practically constant in quantity regardless of the cellular activity, but the quantity varies directly with the *ploidy* (number of genomes) in the nucleus; (4) the configuration of the DNA molecule is suitable for self-duplication, and is at the same time capable of the enormous variability that must be demanded of a genetically active substance; (5) DNA shares with proteins the distinction of occurring in all living things (except some RNA viruses); (6) demonstrable changes in bacteria (transformation and transduction) have been proved to come about as a result of experimental interference with their DNA; and (7) virus particles send only their nucleic acid cores, not their protein coats, into a host cell before virus replication.

CHROMOSOMES

The Structure of Chromosomes.   Chromosomes consist mainly of proteins and nucleic acids. The proteins are simple ones of relatively small size, and are mostly of two general sorts: *histones* containing the amino acids lysine and arginine, and *non-histones* with tryptophan and, of course, other amino acids. The proteins seem to furnish the structural framework of chromosomes and to be involved in metabolic activity. They vary in amount in different tissues and at different times in the same organism. The nucleic acids have been investigated steadily for nearly a century, and intensively since about 1950.

The Nucleic Acids. Although DNA is spoken of as a compound, like hydrochloric acid, it is more properly a class of compounds resembling one another in shape and in the residues which they yield when degraded. The residues are combined in as many as there are, or have been, or ever will be, individual plants and animals (some non-sexual organisms excepted).

Biochemical analysis of the residues, in connection with X-ray diffraction studies, indicated that DNA is, architecturally speaking, a two-stranded coil, with sub-units connecting the strands. A five-carbon sugar (deoxyribose), phosphoric acid, and four different nitrogenous compounds can be obtained from most samples of DNA. Two of these compounds, thymine (a pyrimidine) and adenine (a purine), are regularly found in equi-molecular amounts; and the same is true of the other two bases: cytosine (a pyrimidine) and guanine (a purine). These facts led to the conclusion that in the double-stranded DNA molecule, thymine must be paired with adenine, and guanine with cytosine.

Fig. 13/1 is a simple, schematic view of a piece of a DNA molecule with its components approximately in position, although for the sake of clarity, liberties have been taken with the proportions of the atomic arrangements. A portion of a DNA molecule, unwound, flattened out, and marked with labels, might look somewhat like this:

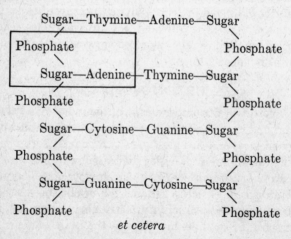

The rectangle encloses a portion, a *nucleotide*, consisting of a nitrogen base, a sugar, and phosphate.

Ribonucleic acid (RNA) resembles DNA in that it, too, occurs in chromosomes, in mitochondria, and in plastids, and is a long molecule consisting of a chain of nucleotides. Both absorb ultraviolet light at a wave-length of 2,600 Å, but RNA differs from DNA in several ways: in RNA, the thymine of DNA is replaced by a different base, uracil; the RNA molecule is single-stranded instead of double; the sugar in RNA, ribose, has one more oxygen atom than the deoxyribose of DNA. The specific enzyme, deoxyribonuclease (DN-ase) hydrolyses DNA; RN-ase hydrolyses RNA. Only DNA gives a positive Feulgen reaction, whereby the decolorized dye, basic fuchsin, becomes cherry red.

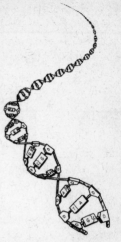

Fig. 13/1.   Fragment of a DNA molecule, showing arrangement of the phosphate, deoxyribose, and nitrogen bases.

DNA FUNCTION

The two functions of DNA are: (1) self replication, which makes possible the repeated division of the chromosomes and consequently of the cells; and (2) control of the synthesis of the structural proteins and of the enzymes which regulate cellular activity.

Replication.   A number of experiments show that the production of two identical DNA molecules from one molecule involves the separation of the two strands of the original double coil, followed by the rebuilding of a complete double-stranded molecule from each "half," using the old "half" as a mold. The mechanics of the methods involved are indicated in the scheme on the following page.

At first, we see the original, or "parent," molecule. At the beginning of the duplication process, the two strands begin to uncoil and the chemical bonds between the organic bases break in sequence. After the bonds are broken, the two strands can separate.

Above, right, the two nucleotide strands begin to separate.

Each strand is furnished with nucleotides from the cytoplasm. When the proper nucleotides have been put into position—and only the proper ones will "fit"—the two molecules each consist of a new and an old strand. When the two new DNA molecules are complete, they are exactly like each other, and like the parent molecule.

Protein Synthesis. A strand of DNA not only produces new molecules exactly like itself, but also acts as a mold for the synthesis of the slightly different strands of RNA. Different kinds of RNA, enzymes for each activity, amino acids, ribosomes, energy-sources, and probably some still unknown factors, all cooperate in the production of proteins. Protein formation is critical in the activity of any cell not only because proteins are part of the structure of cells but also because the total behavior of a cell is essentially an expression of the action of its particular enzymes, and enzymes are proteins.

In the production of a complete protein molecule from its component amino acids (see Chap. 11), the cell must provide the same essentials as would be required for any synthetic process:

The materials for the build-up, in this case the amino acids.

An energy source to drive the process to completion, in this case ATP and the related guanine triphosphate (GTP).

An organized system, in this case the nucleic acids, the ribosomes, and the specific enzymes which catalyze each step in the process.

If any single factor is lacking or disarranged, the entire system either stops or produces an abnormal final product. How such a series of interrelated actions came into being during the evolution of cells is unknown, but it is believed to be of great age because of the similarity between the functional components in cells of such widely separate organisms as bacteria and the most complex higher plants and animals.

Several critical technical points must be understood if the final production of protein molecules is to be understood.

THE STRUCTURE OF DNA. DNA mediates the synthesis of compatible strands of RNA, whose molecules are different from the "parent" DNA. The structural and chemical differences between DNA and RNA have been listed earlier; here the concern is with functional differences. Presumably the double-stranded DNA can unwind and open up, breaking the relatively weak hydrogen bonds which hold purine to pyrimidine in the intact DNA molecule. When RNA is to be made, the proper nucleotides are brought into position. For example, in an opened DNA strand, a thymine-containing *deoxyribose* nucleotide could put into position an adenine-containing *ribose* nucleotide; then if the adjacent deoxyribose nucleotide contained guanine, the corresponding cytosine-ribose nucleotide would be attached to the previously attracted adenine-ribose nucleotide, and so on until a new RNA strand is formed. Thus the arrangement of the nucleotide bases in RNA, a critical factor in protein synthesis and in the life of a cell, is determined by the arrangement of bases in DNA.

THE DISTRIBUTION OF RNA. In a cell nucleus, the DNA mediates the production of three kinds of RNA which pass from the nucleus to the cytoplasm. The three kinds of RNA are *ribosomal RNA, messenger RNA*, and *transfer RNA*.

Ribosomal RNA—is bound with proteins in discrete granules in the cytoplasm, the ribosomes (Chap. 4). It furnishes a place for activity of the other kinds of RNA, but its exact action in protein synthesis is not known.

Messenger RNA—(also called m-RNA, *informational RNA,* or *template* RNA) unites the ribosomes in strands. A row of ribosomes strung along a strand of messenger RNA is a *polyribosome,* or simply a *polysome*. It is the site of protein synthesis. Although the ultimate plan for amino acid arrangement is in nuclear DNA, the immediate plan, the "on-the-spot blueprint," is the messenger RNA, in which the sequence of nucleotides directly regulates which amino acid goes where, and consequently which protein will be built.

With four bases (adenine, cytosine, guanine, uracil) available to make nucleotides, and 20 amino acids needing to be designated, the RNA must use at least three bases at a time in order to provide 20 different combinations. One base at a time could designate only four different things; two at a time could designate sixteen; but three at a time (4^3) can designate 64, or more than the needed 20. In a living cell, three bases in a particular sequence can specify the placement of one particular amino acid. A bit of messenger RNA having three bases is a *triplet,* and its precise arrangement is determined by a triplet of bases in the parent DNA, the DNA triplet constituting one *codon* because it contains the code for a specific amino acid.

Transfer RNA—(also called t-RNA, *soluble RNA,* or *adapter RNA*) consists of relatively short strands of only about 75 nucleotides. Each kind of amino acid can be attached to one kind of t-RNA, depending upon its base sequence. Besides, each molecule of t-RNA has a triplet of bases complementary to a triplet on the m-RNA strand which it will meet at the site of the polysomes. This ability of the t-RNA to "mate" at some exact site with m-RNA determines the position of the amino acid when it is incorporated into a polypeptide chain (protein).

A SUPPLY OF AMINO ACIDS.   These must be available in the cytoplasm, having been converted, via many pathways, from their ultimate source in photosynthesis. Higher plants, and most simpler ones, make their own amino acids from nitrates and sugars, but animals and some lower plants must obtain some or all of their preformed amino acids from an outside source.

A SYSTEM OF ENZYMES.   Another essential factor for protein synthesis is that there must be an adequate system of enzymes in order to catalyze the reaction of each amino acid. Thus, if 20 different amino acids are to be incorporated into a protein, the cell must provide a specific enzyme for each of them. Inasmuch as each enzyme is itself a protein which had to be synthesized by some cell, the whole process of protein production is an endless array of biochemical generations extending back in time to an unknown beginning.

AN ENERGY SUPPLY.   This is necessary because the synthesis of a large, condensed, highly organized protein molecule from its individual components is an *endergonic* (or endothermic) reac-

tion, that is, an energy-requiring one which cannot proceed unless energy in the form of chemical bonds is put into the final
product. As in most cellular activities, energy is made available
in the form of the phosphate bonds of ATP. When the bonds are
broken, liberating the phosphate and ADP, the energy of the
bonds can be transferred to some other molecule. In protein synthesis, the energy is used to activate an amino acid which can,
when so activated, join up with other amino acids by peptide
linkages. In a manner not now understood, another energy source,
GTP, is also required in the final steps of amino acid condensation.

A SUITABLE ENVIRONMENT. Protein synthesis can only occur
in a suitable environment which must exist either in a living cell
or in a properly prepared synthetic medium. Certain conditions,
many of them still incompletely defined, must be maintained,
especially with respect to temperature, pH, ionic balance, and
water content.

The Sequence of Events in Protein Synthesis. The mechanism through which amino acids are incorporated into proteins has
been investigated primarily in bacteria, but it is believed to be
essentially the same in all organisms. The steps, in brief, are as
follows:

1. Nuclear DNA, catalyzed by RNA synthetase, mediates the
production of ribosomal RNA which goes out from the chromosome to become incorporated in ribosomes. DNA also controls
production of m-RNA which also goes to the ribosomes and binds
them together in strands, and of t-RNA which goes into the cytoplasm.

2. An amino acid in the cytoplasm is activated by reaction
with ATP and the specific enzyme for that amino acid, with the
liberation of phosphate, and the production of an enzyme—amino-
acid—adenylic-acid complex.

3. The amino acid complex transfers the amino acid to a
specific t-RNA molecule. The enzyme and the adenylic acid are
released, and a new complex of activated amino acid and t-RNA
results. Each kind of amino acid not only has its own enzymes,
but also its own t-RNA.

4. The new complex goes to a ribosome, where a specific triplet
of the t-RNA, with its attached, activated amino acid, meets a

complementary triplet of the m-RNA which is associated with the ribosomes. The only place where a particular t-RNA can fit is that place on the m-RNA strand.

5. The amino acid is thus lined up on the ribosome in such a way that it takes its place with other amino acids in a growing polypeptide chain. This activity requires an enzyme (probably one which promotes peptide bonding) and GTP.

6. The t-RNA is released to make another amino-acid-carrying trip. The newly-formed strand of united amino acids, each in its proper place on the sequence, is liberated from the ribosome, with the free amino group of the first amino acid at the starting end and the free carboxyl group of the last amino acid at the finish. The release of this large new molecule is probably brought about by another enzyme.

7. After the new protein is released in the form of a linear chain of amino acids joined by peptide bonds, it may be variously twisted, coiled, bent, or otherwise convoluted until it achieves its final form, with cross bonding especially between sulfur-containing amino acids. How the ultimate geometry of the protein molecule can be achieved is not known.

THE GENETIC CODE

By using only uracil as a building nucleotide, a kind of RNA has been synthesized which is called poly-uracil or simply poly-U. The triplets in poly-U are, of course, all UUU. When poly-U is mixed with all the amino acids, plus their activating enzymes and the other requisites for polypeptide synthesis (ribosomes, ATP, etc.), only the amino acid phenylalanine is incorporated into a chain of poly-phenylalanine. This indicates that UUU is the code triplet for phenylalanine. Similarly, poly-adenine or poly-A (with triplets of AAA) is the code for the amino acid lysine, and CCC is the code for proline. Combinations of nucleotides code for other amino acids, but the degree of confidence in our knowledge of them is generally less than for poly-A, poly-U, and poly-C.

Amino acids and the triplets that are thought to code for them are given in the following columns. A number of the amino acids can apparently be coded for by more than one triplet; this multiple coding is known as *"degeneracy."* Some triplets do not code for any amino acid; such triplets are thought to act as signals for the end of a polypeptide chain, and are called "punctuation signals."

alanine   GCA? (maybe),
   GCG?, GCC?, GCU
arginine   CGA, CGG?, CGC,
   CGU?, AGA, AGG
asparagine   AAC, AAU
aspartic acid   GAC, GAU
cysteine   UGC, UGU
glutamic acid   GAA, GAG
glutamine   CAA, CAG
glycine   GGA, GGG?, GGC?,
   GGU
histidine   CAC, CAU
isoleucine   AUC, AUU, AUA?
leucine   CUA, CUG, CUC,
   CUU, UUA, UUG

lysine   AAA, AAG
methionine   AUG
phenylalanine   UUC, UUU
proline   CCA, CCG?, CCC,
   CCU
serine   AGC, AGU, UCA,
   UCG, UCC, UCU
threonine   ACA, ACG, ACC,
   ACU
tryptophan   UGA, UGC
tyrosine   UAC, UAU
valine   GUA?, GUG, GUC?,
   GUU

"punctuation"   UAA, UAG

FEED-BACK CONTROL

Presumably every cell in an organism has an entire set of chromosomes in its nucleus, and is theoretically capable of performing any activity. Yet, obviously, no single cell in a complex multicellular organism does perform all the syntheses which its DNA could direct. Some means must exist for the determination of which code triplets will function, and when they will do so. Exactly how DNA segments are "turned on" and "turned off" is not known, but some clues are available.

The regulation of cellular activity is in general a negative feed-back. A simple example of a *negative feed-back* system can be found in human clothes: on a *cold* day, a man dons a *warm* coat, but if the weather turns *warmer*, the man changes to *less warm* clothing. Similarly, if a cell synthesizes an *excess* of some compound, the very presence of the excess can cause synthesis to *slow down or stop;* or *insufficient quantities* of the compound may cause synthesis to *start or speed up.*

An example of cellular feed-back is found in the regulation of production of the purines and pyrimidines which are incorporated in equal amounts in DNA. The synthesis of more thymine (a pyrimidine) than adenine (a purine), for instance, would be inefficient because the thymine would have nothing to pair with during DNA synthesis, and cells generally cannot afford to be inefficient. Experimentally, it can be shown that a cell provided with an *excess* of thymine will *curtail* thymine production and increase adenine production. Other examples can be drawn from studies on amino acid synthesis. Presumably similar systems are at work in the thousands of enzyme systems in a cell, and information is fed back to some starting point in such a way as to regulate efficiently the total behavior of the cell.

HOMEOSTASIS

One manifestation of all the foregoing activities (DNA action, the various syntheses, and feed-back control) is the ability of a cell to maintain itself in a reasonably steady condition. It does this, not by remaining inactive, but by keeping itself in a constant state of activity, eliminating some materials, bringing in new ones, replacing some molecules with other ones, and keeping up an endless turn-over of its own parts. Experiments with amino acids labeled with radioactive carbon 14 show that even the structural proteins of an active cell are not permanently built into the framework of the cell, but are continually being replaced, with the result that even a mature cell in an apparently stable condition is really maintaining itself, by its own activity, in a state of *homeostasis*.

Chapter 14

Growth: Environmental Effects

What an organism *can* become is determined primarily by the DNA in its cells; but what it *does* become is a result of the action of the organism's environment on the entire organism, acting in conjunction with the genetic capabilities of the organism. Given optimum conditions for apple culture, no peach seed will grow into an apple tree. At the same time, an apple seedling will not develop into a productive tree if the temperature is too high and the soil moisture too low, or other environmental conditions are unfavorable.

Biological growth, movement, and morphological differentiation are frequently described as reactions, or *responses*, to external environmental forces, or *stimuli*. The most common, biologically effective stimuli are: aspects of gravitational pull; temperature variation; mechanical pressure; presence of substances in solution; and light, which varies in quality (wavelength of radiation), intensity (energy per unit area per unit time), and duration.

CHEMICAL REGULATORS

External influences usually bring about their perceptible effects by means of some intermediate regulatory compound. The best-known regulators are the *auxins, giberellins, kinins,* and *phytochrome.*

Auxins. The best-known of the hormone-like substances that influence plant growth is an auxin, indole acetic acid, with this chemical structure:

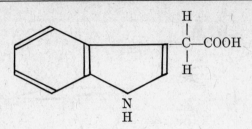

The following facts about auxin are experimentally demonstrable:

1. Auxin is produced principally in the growing shoot apex of a plant.

2. It moves downward under the influence of gravity.

3. It moves away from its region of synthesis by diffusion or by directional transport.

4. It is driven by unidirectional light from an illuminated side to a darker side of a plant.

5. A very low concentration of auxin in a stem stimulates cell elongation, but the same concentration in a root can inhibit cell elongation.

These facts can be used to explain the bending movements of plants in response to gravity and unilateral illumination. For example, if a young plant begins by chance to grow horizontally, the auxin moves from the apex where it is produced, toward the root and at the same time accumulates more on the lower side than on the upper side of the young stem. Continued transport of the auxin results in its accumulation along the lower side of the root. Since auxin stimulates cell elongation in the stem, the lower side of the stem grows faster than the upper side, thus bending the stem upward. Since auxin inhibits cell elongation in the root, the upper side of the root consequently grows faster than the lower side, thus bending the root downward (Fig. 14/1).

Even if the stem or root deviates only slightly from a vertical position, the phenomena described above will take place, and the primary shoot will grow upward and the primary root downward. These effects work only on the primary organs, especially while a young seedling is becoming established in the soil; secondary or branch shoots and roots show less sensitivity or none.

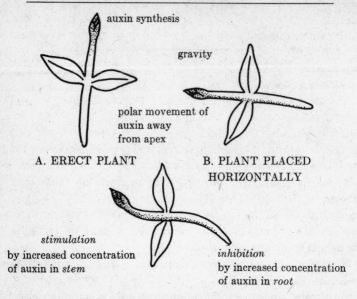

auxin synthesis

gravity

polar movement of
auxin away
from apex

A. ERECT PLANT

B. PLANT PLACED
HORIZONTALLY

stimulation
by increased concentration
of auxin in *stem*

inhibition
by increased concentration
of auxin in *root*

C. PLANT GROWING DIFFERENTIALLY AFTER
REDISTRIBUTION OF AUXIN

Fig. 14/1. Representation of the auxin regulation of geotropic response. A seedling, having been laid down, points its apex away from the pull of gravity (negative geotropism) and its primary root toward the earth's center (positive geotropism).

Tropisms. A growth response which a plant makes with respect to a directional stimulus is a *tropism*. The bending of a shoot away from the pull of gravity is *negative geotropism;* the bending of a root toward the center of the earth is *positive geotropism*.

Bending of stems toward light, *positive phototropism*, is caused by the accumulation of auxin on the dim side of the stem, with faster cell elongation there, and consequent bending. Roots show a weak *negative phototropism*.

Auxin affects cell division and maturation as well as elongation. Lateral buds are held in dormancy by sufficient concentrations of auxin, as is shown by the phenomenon of *apical dominance*. In some plants, as in many coniferous trees with a main central trunk, the lateral buds near the apex do not begin to break out as long as the apical bud is intact; but if the apex is removed, the lateral buds can begin to grow. Apical dominance can be artificially maintained by applying auxin to the cut stump after an apex has been removed. Auxin can also stimulate the initiation of a new root growth, and can influence the maturation of xylem cells in stems.

The precise mechanism by means of which cell division, elongation, and differentiation is effected in plants is still not known.

Gibberellins. The gibberellins are effective in stimulating elongation of internodes, in altering flowering times, in stimulating fruit development and growth of lateral buds, and in increasing cell divisions. Originally found in parasitic fungi (*Gibberella*), gibberellins have been demonstrated in many various plants and are probably of natural occurrence in most, if not all, plant groups. Their most dramatic action is in making genetically dwarf plants grow to normal size.

Kinins. The kinins, or kinetins, can stimulate cell divisions, cause lateral buds to grow, inhibit growth of adventitious roots, and increase leaf growth. They are thus in some ways antagonistic to the auxins, but in entire plants the combination of hormonal compounds is effective in maintaining a balance of growth, enlargement, and differentiation. Kinins, originally found in yeast, have been isolated from many kinds of plant cells, and are probably of universal occurrence in plants.

Phytochrome. In addition to its energy-transferring effects in photosynthesis, light also has profound effects upon the developmental activities of plants, and controls many aspects of their formative, or *morphogenetic*, features. One of the most intensively studied of the light sensitive chemical compounds in plants is *phytochrome*, a still incompletely characterized protein, bluish in color, and capable of regulating plant growth in a variety of ways.

Phytochrome exists in plants in two forms which may be converted from one form to the other, either "spontaneously," that is, via a mechanism not yet understood, or by the action of specific wave-lengths of light. In one form, known as P_{660} or P_R (phytochrome "red"), the greatest absorption of light is in the red region of the spectrum, at about 660 millimicrons, 6600 Å, or 660 nm (nanometers). When exposed to light of such wavelength, P_R is converted to P_{FR} (phytochrome "far red," or P_{730}).

P_{FR}, absorbing maximally at 730 nm, is converted by such radiation back to P_R.

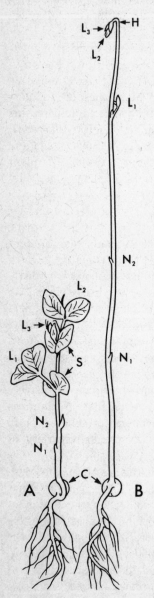

Daylight provides more of the red than the far-red, and consequently daylight acts as essentially a red source as far as phytochrome is concerned.

Phytochrome is known to affect the activities of plants in a number of ways, although the exact mechanism is not understood. Phytochrome activity is not related to the other light-induced activities of chlorophyll synthesis, photosynthesis, and phototropism.

The best-known phytochrome actions are in (1) *germination*, (2) *control of etiolation*, and (3) *photoperiodism*.

GERMINATION. Some varieties of seeds (notably lettuce of the Grand Rapids variety) will not germinate unless they receive a minimal dose of red light. Daylight contains enough red to cause germination. If kept in the dark or irradiated with far-red,

Fig. 14/2. Pea seedlings grown for 9 days: A, in light; B, in darkness. Notice extreme differences in leaf expansion and in stem elongation (note positions of successive leaves L_1, L_2, L_3, and of nodes N_1 and N_2 at which only scale leaves are formed). C, cotyledons. S, stipules. H, hook-shaped part of growing region of stem (note absence of hook in A). (Magnification ½ ×.)*

* From *The Living Plant* by Peter M. Ray. Copyright © 1963 by Holt, Rinehart and Winston, Inc. Reproduced by permission of Holt, Rinehart and Winston, Inc.

the seeds remain dormant. When treated with alternating red and far-red, the seeds respond to the last treatment. For example, if the last light provided is red, the phytochrome is converted to P_{FR} and the seeds can germinate; but if the final treatment is far-red, the phytochrome is converted to P_R and the seeds do not germinate.

CONTROL OF ETIOLATION. When grown in the dark or in far-red light, seedlings become *etiolated*, that is, their leaves do not expand, the internodes elongate excessively, the chloroplasts do not develop chlorophyll, and a sharp bend in the stem just below the growing point keeps its curvature. (Fig. 14/2). Seedlings grown in daylight are "normal"; that is, the internodes elongate only enough to separate the nodes, the leaves expand, the plants become green, the shoot tip straightens, and other changes occur. These features, except for the greening of chloroplasts, are regulated by phytochrome, which follows the red—far-red pattern. When a series of seedlings is grown in a spectrum of controlled wave-lengths, those seedlings grown in far-red (730 nm) are etiolated, while those grown in red light (below 700 nm) are normal except for their color. From such trials, the responses of an organism to a spectrum can be observed, and the result is called an "action spectrum," in comparison with the "absorption spectrum" obtained by measuring the amount of light at various wave-lengths absorbed by a chemical compound. The similarity between the absorption spectra of phytochromes and the action spectrum of etiolation strengthens the idea that phytochrome is involved in the morphogenesis of seedlings.

PHOTOPERIODISM. Some plants (cockleburs, some varieties of tobacco, soybeans, and chrysanthemums) make flowers only when days are short in comparison to the length of nights. Other plants (rose mallows, spinach, some grasses) flower only during long days. Still others (tomatoes, corn, buckwheat) are "day-neutral," and flower without regard to day length.

Short-day plants—will flower if they receive the following treatments: long, uninterrupted dark periods; or long dark periods interrupted by a short light period, provided the light "flash" is followed by a "flash" of far-red light.

Short-day plants will remain vegetative if they receive: short dark periods; or long dark periods which are interrupted midway by a short light "flash."

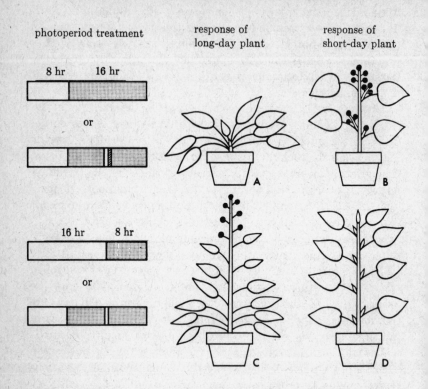

photoperiod treatment   response of   response of
                               long-day plant   short-day plant

8 hr   16 hr

or

16 hr   8 hr

or

Fig. 14/3. Responses of typical long- and short-day plants to short days (above), to long days (below), to short day with interrupted night (below), and with night interruption followed by far red (above, far-red period diagonally shaded). Black circles represent flowers; open symbols show leaves and vegetative buds.

Long-day plants—will flower if they receive: short dark periods; or long dark periods which are interrupted by a "flash" of light.

Long-day plants remain vegetative if they receive: long uninterrupted dark periods; or long dark periods interrupted by a short light "flash," provided the light is followed by a "flash" of far-red light.

Photoperiodic action of flowering follows the pattern of phytochrome reversibility. Phytochrome, by its alteration in light and darkness, provides the plant with a "time sensing" device, inasmuch as the spontaneous reversal of P_{FR} to P_R in the dark proceeds at a rate which is independent of temperature, and permits a plant to flower or to avoid flowering in accordance with the daylength or time of year.

Rhythms which are internally set, as the P_{FR}—P_R conversion, are known as *endogenous rhythms*. Since many of these rhythms are associated with the earth's twenty-four hour cycle, they are called *circadian rhythms*, from Latin "circa diem" (about a day). Some movements of leaves, for example, show daily rhythms even in the absence of normally occurring, external changes in light and temperature. Endogenous rhythms show that plants possess time-measuring systems which are called *biological clocks*. The exact mechanism which functions in these clocks is unknown.

OTHER PLANT MOVEMENTS

Tropisms. Tropisms, other than geo- and phototropism, are *thigmotropism* and *hydrotropism*. Thigmotropism, a response to touch, is demonstrable in tendrils (a growth reaction) and in special leaves, such as the Venus fly-trap and the sensitive Mimosa (a movement caused by turgor change). Phytochrome is thought to act in thigmotropic actions. Hydrotropism, reaction to water concentrations, is so weak that its existence is doubted by many botanists.

Nastic movements. These are movements of leaves and petals, caused by unequal growth of either the upper layers of an organ growing faster than the lower, causing the organ to bend downward (epinasty), or vice versa (hyponasty). No satisfactory explanation for nastic movements is available.

Taxic Responses. These are movements of entire plants, such as motile cells of algae, fungi, sperm cells, etc. The movement may be toward a stimulus (positive) or away from it (negative). Common taxic responses are *phototaxis, chemotaxis, thermotaxis,* and *rheotaxis* (response to electric current).

Chapter 15

Flower Structure and Activities

A flower is not a single organ, but is a branch bearing leaflike and stemlike parts on a short axis. The onset of flowering is determined by heredity, photoperiods, food reserves, and often critical temperatures. Before the onset of flowering, a plant is thought to synthesize a flower-initiating hormone which has been called *florigen* even though the presence of this hormone has never been proved.

Flowers, like other types of twigs, develop from buds. Flowers develop from flower-buds (e.g., in morning-glories, roses) or from mixed buds (e.g., in buckeye). Floral organs develop as protuberances from the growing tip of a bud in basipetal order, i.e., from the apex downward. The tip of a floral twig does not elongate as much as the tip of a vegetative twig. As a result, the floral organs are crowded at the apex of the twig, and not distributed along the twig as leaves are.

THE PARTS OF A COMPLETE FLOWER

A complete flower bears four kinds of floral organs (Fig. 15/1). The tip of the floral twig to which these organs are attached is the *receptacle*. The four organs are *sepals, petals, stamens,* and *pistil.*

Sepals. This outermost circle of leaves, known collectively as the *calyx,* is usually green in color, or is sometimes the same color as the petals. Sepals protect the inner parts of the flower in the bud.

Petals. This circle of organs lies inside the sepals, and is called collectively the *corolla.* They are frequently brightly colored, and often secrete aromatic substances and *nectar* (concentrated sugar solution). Petals attract insects, which are necessary

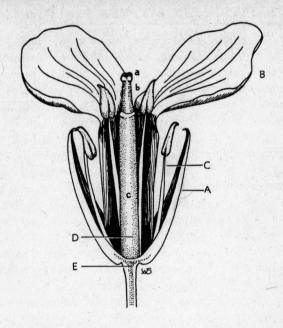

Fig. 15/1. A complete flower. A. Sepal. B. Petal. C. Stamen (anther and filament). D. Pistil; a stigma, b. style, c. ovary. E. Receptacle.

for pollination of many flowers. The numbers of sepals and of petals are usually the same in the same species.

Stamens. These floral organs are situated inside the petals. A stamen consists of a stalk (*filament*), with a pollen-bearing *anther* at its apex.

Pistil. A structure in the center of the flower, the pistil is composed of one organ (*simple pistil*) or of several fused organs (*compound pistil*). An ovule-bearing organ is a *carpel*. A pistil has a base (*ovary*), with a stalk (*style*) arising from the ovary, and a slight enlargement (*stigma*) at the top of the style. Within the ovary the undeveloped seeds (*ovules*) are produced. The ovules are attached to *placentae* inside the ovary.

The sepals and petals are known as *accessory parts,* since they are not directly concerned with reproduction. The stamens and pistil(s) are the *essential parts* of a flower.

VARIATIONS IN FLOWER STRUCTURE

Flowers differ from one another in many ways. From comparative studies of living and fossil types, and from consideration of varying complexities, botanists have tentatively concluded that floral variations offer evidence of the degree of advancement of a species, and tell something of its evolutionary relationships with other species. In the following list of variations, the more primitive (or ancient) condition is given first, followed by the more advanced (or modern, or derived) condition. Generally, but not necessarily, the more primitive condition is structurally simpler than the modern condition. Evolution has generally been toward complexity; but when a simpler condition (as in imperfect flowers) is usually accompanied by a high degree of complexity, then the structurally simpler condition may be considered advanced. The most obvious variations—size, shape, color—are thought to be relatively unimportant in evolutionary schemes.

1. A *complete flower* (e.g., rose) has four kinds of flower parts. An *incomplete flower* lacks one or more of these kinds of parts (e.g., elm, wheat).

2. A *perfect* flower has both stamens and pistil (e.g., rose). An *imperfect* flower has stamens or pistil, but not both (e.g., willow, corn). A *monoecious* plant has stamen-bearing and pistil-bearing flowers on the same plant (e.g., corn). A *dioecious* plant has staminate flowers on one plant, *pistillate* on another (e.g., willow).

3. The numbers of flower parts vary. Dicots have their flower parts in fives, fours, or twos, less frequently in threes. Monocots usually have their flower parts in threes or in multiples of three.

4. Flower parts may be completely separate, or they may be fused in varying degree. Floral organs of the same kind may be fused together (*connation*), or to the floral organs of another type (*adnation*).

5. Flowers with *radial symmetry*, called *regular* flowers (e.g., roses and tulips), are built on a wheel-like basis. Flowers with *bilateral symmetry*, called *irregular flowers* (e.g., snapdragons and orchids), can be divided along a single plane to produce two halves which are mirror images of each other.

6. In *hypogynous* flowers (e.g., tulip), the sepals, petals, and stamens are attached to the receptacle under the ovary, which is

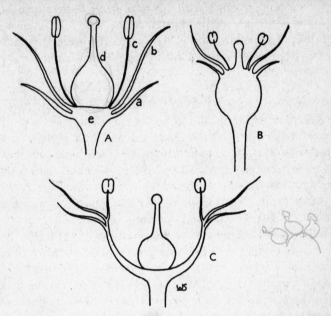

Fig. 15/2. Ovary position in flowers. A. Superior ovary; a. sepal, b. petal, c stamen, d. pistil, e. receptacle. B. Inferior ovary. C. Perigynous flower.

said to be *superior*. In *perigynous* flowers (e.g., cherry), the pistil is in the bottom of a concave receptacle to the edges of which the sepals, stamens, and petals are attached; in such a flower, the ovary may be *superior* or *half-inferior* (Fig. 15/2). In *epigynous* flowers (e.g., honeysuckle), the pistil is sunken into the receptacle, with sepals, petals, and stamens attached above the ovary, which is termed *inferior*.

7. The parts of a flower may be produced on the receptacles in *spirals*, (as in tulip poplars and magnolias); or they may be produced in circles or *whorls* (as in apples).

8. Flowers may be borne *singly* (tulips, magnolias) or in *clusters* (*inflorescences*), as in snapdragons or composites. In composites (e.g., sunflowers) the individual flowers are minute and are borne in large numbers on a flattened disc; the central flowers (*disc-flowers*) of the disc have small radial corollas, the marginal flowers (*ray-flowers*) have large, bilateral corollas. Such

an inflorescence comprising numerous small flowers is called a *head* and is found also in daisies, asters, chrysanthemums, etc. The stalk of an inflorescence, from which the *pedicels* of individual flowers branch, is the *peduncle*. In solitary flowers, the stalk is the peduncle.

POLLINATION

Pollination is the transfer of pollen from a stamen to a stigma. It is brought about by wind, water, and by animals—insects, birds, bats, etc. Most important agents are wind and insects. *Self-pollination* is the transfer of pollen from the stamen to the stigma of the same flower or to the stigma of another flower on the same plant. *Cross-pollination* is the transfer of pollen from an anther to a stigma on another plant.

Insect and wind pollinated flowers differ structurally. Insect pollinated flowers have conspicuous petals, usually produce odors or nectar or both, have small or moderate-sized stigmas, and moderate amounts of pollen, which is often sticky; wind pollinated flowers lack conspicuous petals, are usually odorless and nectarless, have large, flat, or hairy stigmas, and large amounts of light, dry pollen (willows, grasses).

Many kinds of plants have various devices to ensure cross-pollination. Some of these devices are:

1. Imperfect flowers.

2. Difference in time of maturation of stamens and stigma of the same flower.

3. Chemical incompatibility between stigma and pollen. Pollen grains will not germinate on stigmas on the same plant, or the pollen tube cannot grow to reach the ovule, or the young embryo fails to develop.

4. Specialized structural devices—spring arrangements (e.g., *Salvia*), peculiarities of style and stamen structure (e.g., primrose, etc.)

THE DEVELOPMENT OF
POLLEN GRAINS AND OVULES

Pollen Grains. An anther usually contains four masses of *sporogenous tissue*, each mass being the forerunner of an anther sac. The diploid cells of the sporogenous tissue are the *microspore* *mother cells.* They undergo meiosis (Chap. 12), with the result that a quartet or *tetrad* of cells is produced, each being a *microspore*. The microspores separate from one another, each one forms a thick wall around itself, and its haploid nucleus undergoes a mitotic division. The final product is a *pollen grain* with two nuclei. One nucleus, the *tube nucleus*, serves no known function and is presumably a remnant of earlier times in the evolution of flowering plants. The second nucleus, the *generative nucleus*, is destined to undergo another mitotic division later to produce two sperm nuclei (Fig. 15/3).

Pollen grains are variously colored (orange, yellow, red, brown, purple) and variously sculptured with pores, spines, streaks, and indentations. The morphology of pollen is so distinctive for each kind of plant that fossilized pollen from ancient forests can be identified, and the old flora (and consequently the old climates) from the geological past can be reconstructed.

Ovules. An ovary may be *simple* (composed of one element or *carpel*, as in peas and beans) or *compound* (composed of

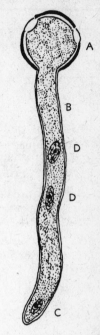

Fig. 15/3. Pollen grain with tube. A. Pollen grain. B. Pollen tube. C. Tube nucleus. D. Sperm nuclei.

more than one carpel, as in oranges). Cavities in the ovary (*locules*) contain protruding masses of tissue (the *ovules*), which will develop into seeds. Each ovule is attached by a stalk (*funiculus*) to the ovary at a region called the *placenta* (Fig. 15/4).

In each ovule, one cell is conspicuously different from the surrounding cells, usually much larger. This cell is the *megaspore mother cell*. It undergoes meiosis, producing four haploid *megaspores* in a row. Three of these megaspores usually abort, while the fourth megaspore enlarges greatly and then undergoes three mitotic divisions to yield eight haploid nuclei in a fluid-filled *embryo sac*.

An ovule is covered by two layers of cells, the *integuments,* except at one place, usually near the connection of the funiculus. The gap in the integuments is the *micropyle* through which the pollen tube enters the embryo sac. The integuments later become the seed coats. The eight nuclei in an embryo sac are usually disposed as follows: (1) three are near the micropyle, one being the *egg nucleus* or female gamete, the other two (the *synergids*) disintegrate; (2) three nuclei (the *antipodals*) at the end opposite the micropyle also disintegrate; (3) the remaining two (the *polar nuclei*) combine to make a diploid nucleus. As with most biological sequences, this one is subject to great variation, but it is essentially constant in the following features: reduction division in the megaspore mother cell, formation of a haploid egg nucleus, and production of a second nucleus (diploid or otherwise). (Fig. 15/5.)

Fig. 15/4. Cross-section of lily ovary showing three carpels. A. Locule. B. Funiculus. C. Ovule. D. Embryo sac.

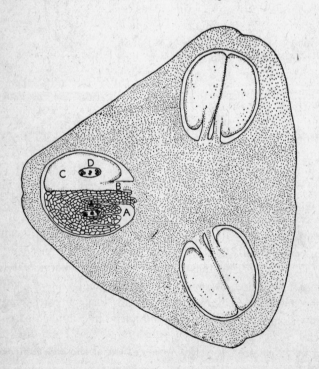

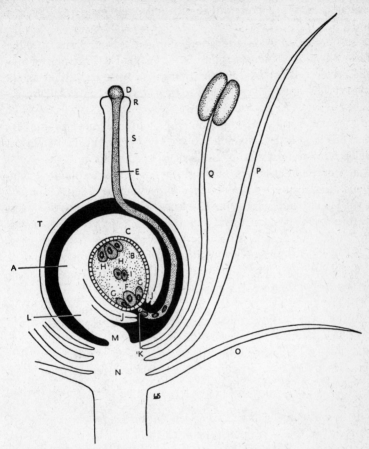

Fig. 15/5. Longitudinal section of flower, showing pollen-tube growth and ovule structure. A. Ovule. B. Embryo sac (megagametophyte). C. Nucellus. D. Pollen grain. E. Pollen tube. F. Egg nucleus. G. Synergids. H. Antipodals. I. Polar nuclei. J. Integuments. K. Pollen tube entering micropyle, with one tube and two sperm nuclei. L. Funiculus. M. Placenta. N. Receptacle. O. Sepal. P. Petal. Q. Stamen. R. Stigma. S. Style. T. Ovary.

POLLEN TUBE GROWTH, FERTILIZATION AND SEED DEVELOPMENT

Following the landing of a pollen grain on a stigma, which is often covered by a sticky fluid, hairs, or roughened protuberances which hold the pollen grains, the following incidents occur in order:

1. The pollen grain swells and germinates, and forms a *pollen tube* which, growing down through the style by digesting some of the stylar cells or by growth through a stylar canal, enters the ovary. The growth of the pollen tube may be controlled by the tube nucleus. (Figs. 15/3, 15/5.)

2. A pollen tube enters the micropyle of an ovule in the ovary and discharges into the embryo sac two *sperm nuclei*, which develop from the division of the generative nucleus.

3. One haploid sperm nucleus fuses with the haploid egg nucleus, thus forming a diploid *zygote*, or "fertilized egg."

4. The second sperm nucleus fuses with the two polar nuclei to form the triploid *endosperm nucleus*, the result of triple fusion. This behavior of both sperms is known as *double fertilization*.

5. The tube nucleus, synergids, and antipodals disintegrate.

6. The zygote, by numerous cell divisions, develops into the *embryo* of the seed.

7. The endosperm nucleus develops by numerous cell divisions into the *endosperm* (food storage) tissue of the seed.

8. The integuments become the seed coats of the seed.

9. Following fertilization, the ovary and its ovules increase in size. In a few plants, fruits develop without fertilization (e.g., navel orange, banana, pineapple), a condition known as *parthenocarpy*.

10. Development of fruit is dependent on the presence of auxin, supplied naturally by pollen or artificially by spray.

Chapter 16

Fruits and Seeds

A fruit is a matured *ovary;* a seed is a matured *ovule.* Often fruits have adhering to them other floral parts. Such fruits are called *accessory fruits.*

FRUITS

The cavities of a fruit, within which the seeds are produced, are *locules.* The wall of a ripened ovary is the *pericarp.* The pericarp consists of three layers of tissue which are not always distinguishable: the outermost wall, or *exocarp,* is usually only one cell thick; the *mesocarp,* or middle wall, is thicker than the exocarp and contains the conducting tissues; the *endocarp,* or innermost tissue, surrounds the locules. (Fig. 16/1.)

Fruits are classified as follows:

Simple Fruits. A simple fruit develops from a single ovary of a single flower.

Fleshy Fruits. These simple fruits are soft and pulpy at maturity.

Berry. The entire pericarp becomes soft and fleshy (e.g., grape, tomato, banana, watermelon, orange).

Drupe. The exocarp and mesocarp are soft and fleshy, the endocarp becomes hard and stony (*pit*). Inside the pit is usually one seed (sometimes two or three), (e.g., peach, cherry).

Dry Fruits. These are dry and hard or papery at maturity. Dry fruits are of two kinds—dehiscent and indehiscent.

Dehiscent Fruits. Split open along one or more definite seams (*sutures*). (1) Capsule. A dry fruit formed from a compound ovary (composed of more than one fused carpels), (e.g., poppy, snapdragon). (2) Legume. Develops from a single carpel, splits

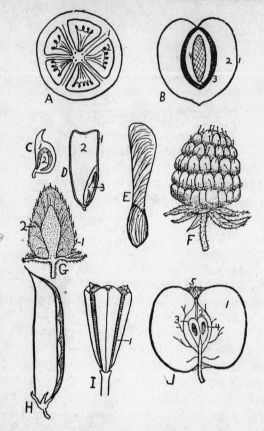

Fig. 16/1. Some common types of fruits. A. Berry (tomato): 1, pericarp; 2, seeds. B. Drupe (plum): 1, exocarp; 2, mesocarp; 3, stony endocarp; 4, seed. C. Achene (buttercup): 1, fruit; 2, seed. D. Caryopsis (corn): 1, fused fruit and seed coats; 2, endosperm of seed; 3, embryo. E. Samara (maple). F. Aggregate fruit (raspberry). G. Accessory fruit of strawberry: 1, achene; 2, receptacle. H. Legume (lima bean). I. Capsule (tulip): 1, seeds. J. Accessory fruit of apple (pome): 1, cortex of receptacle; 2, pith of receptacle; 3, pericarp; 4, seed; 5, remnants of calyx.

along two seams (e.g., pea, bean). (3) Follicle. Develops from a single carpel, splits along one seam (e.g., larkspur, columbine). (4) Silique. Develops from two carpels, which separate at maturity, leaving a partition wall (e.g., mustard).

Indehiscent Fruits. Do not split by definite seams or pores at maturity. (1) Achene. One seed attached to inside of ovary at one point; ovary wall and seed coat separable (e.g., sunflower, buttercup). (2) Caryopsis (grain). One seed, coat of which is fused with ovary wall and not separable from it (e.g., corn, wheat). (3) Samara. One or two seeds. Pericarp has wing-like outgrowths (e.g., ash, maple). (4) Nut. Hard, one-seeded fruit developed from a compound ovary (e.g., acorn, hazelnut). (5) Schizocarp. Carpels usually two, separating at maturity. Each carpel has one seed (e.g., carrot, parsnip).

Aggregate Fruits.   A fruit which develops from separate simple ovaries of a single flower (e.g., blackberries, raspberries).

Multiple Fruits.   A fruit which develops from the ovaries of several flowers borne close together on a common axis (e.g., Osage orange, pineapple, mulberry).

Accessory Fruits.   A fruit in which the major portion consists of tissue other than ovary tissue. Common types are:

1. Apples and pears, in which the true fruits are the walls and locules of the core, and the fleshy portion is the swollen receptacle and calyx surrounding the core. Such a fruit is called a *pome.*

2. Strawberries, in which the true fruits are tiny achenes on the surface of a much enlarged, sweet, fleshy receptacle.

SEED STRUCTURE

A seed consists of a *seed coat,* which develops from the integuments of the ovule; an *embryo,* which develops from a fertilized egg or zygote; and an *endosperm,* a food-storage tissue, which develops from the endosperm nucleus of the embryo sac. In most cases, the embryo of a seed begins to digest and utilize the food stored in the endosperm when the seed is planted. In other seeds (beans, peas) the embryo digests and absorbs the endosperm before the seed leaves its parent plant; in such seeds no endosperm is present at maturity. (Fig. 16/2.)

Seed Coat.   This part of the seed is usually tough and is partly impervious to water. It prevents excessive evaporation of water from inner parts of seed and often prevents entry of parasites. Hard seed coats may prevent mechanical injury. Various structures may be visible on the surface of a seed coat:

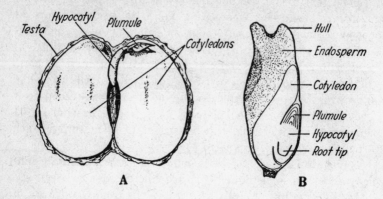

Fig. 16/2.   Seed structure. A. A lima bean laid open to show the parts of the embryo, natural size. B. Longitudinal section of a grain of corn (3 ×).*

HILUM.   A scar left by the breaking of the seed from its stalk.

MICROPYLE.   A small pore near the hilum.

RAPHE.   A ridge on the seed, caused by the bending of the seed against the seed stalk.

Endosperm.   Cells of the endosperm have 3x chromosome number, since they develop from the endosperm nucleus, which is formed by the fusion of three nuclei—two polar nuclei and a sperm nucleus. The endosperm stores food—starch, proteins, oils, etc. Some seeds are valued by man for their starch (wheat), others for protein (beans), others for fats (coconut). The seeds of legumes (beans, peas, clover) have no endosperm at maturity.

Embryo.   The embryo, or miniature plant of the seed, consists of the cotyledon, epicotyl, and hypocotyl.

COTYLEDON(s).   Cotyledons are seed leaves. Seeds of monocotyledons have one, seeds of dicotyledons have two cotyledons. Cotyledons digest and absorb food from the endosperm, or store food.

EPICOTYL.   This is the part of the embryo axis above the point of attachment of the cotyledons. The epicotyl contains meristematic cells which grow into the shoot when the seed sprouts (*ger-*

---

* Reprinted by permission from *Fundamentals of Biology*, by A.W. Haupt, published by the McGraw-Hill Book Company.

minates). The growing tip of the epicotyl is often called the *plumule*.

HYPOCOTYL.   This part of the embryo axis is below the point of attachment of the cotyledons. The meristematic cells of the hypocotyl develop into the primary root when the seed sprouts. The growing tip of the hypocotyl is the *radicle*.

Seeds vary in structure in different kinds of plants. Three common types are the following:

BEAN. A mature bean seed has two large, fleshy cotyledons, no endosperm, and a small embryo axis, with a pair of tiny leaves at the apex of the epicotyl.

CASTOR BEAN.   A mature seed has two flat, thin cotyledons, a short epicotyl and hypocotyl. The embryo is embedded in a large, oily endosperm. A large spongy structure called the *caruncle* lies at one end of the seed.

GRAIN (e.g., corn).   A mature "seed" (caryopsis, or one-seeded fruit) has a large, horny or mealy endosperm (sometimes both), moderate-sized embryo that consists of a shield-shaped cotyledon, an epicotyl covered by a sheath (*coleoptile*), and a hypocotyl covered by a sheath (*coleorhiza*). The seed coat and pericarp are fused.

SEED GERMINATION

In seed germination, a seed takes up water and swells, food is digested, respiration increases, and cell division occurs, following which the embryo grows and the seed coat is ruptured. The hypocotyl is usually the first part of the embryo to emerge from the seed coat; this is advantageous, for the young root can immediately begin to absorb the water and minerals necessary for growth. The epicotyl emerges next. A sprouted embryo is a *seedling*.

Germination Patterns.   The pattern of germination varies in different species, as shown by:

BEAN.   The lower part of the hypocotyl becomes the root. The upper part is crook-shaped and pulls the cotyledons and epicotyl above the soil surface. The crook in the hypocotyl forces an opening through the soil and pulls the cotyledons and epicotyl up through this opening. Cotyledons become temporarily photosynthetic, then dry and fall after giving up their food.

CASTOR BEAN. Similar to bean in that cotyledons and part of hypocotyl rise above soil. Cotyledons are flat and leaf-like, become green, persist longer than those of bean. Some of the endosperm is carried above ground with the cotyledons.

PEAS. Upper part of hypocotyl does not grow above ground; thus, cotyledons remain below soil.

GRAIN (corn). Cotyledon remains in soil. Primary root system soon dies. Adventitious roots develop and form permanent root system.

Conditions affecting Germination. Seed germination is influenced by various external and internal conditions.

EXTERNAL CONDITIONS.

Moisture. Seeds must have abundant moisture to germinate. Excessive moisture may cause rotting if oxygen is excluded. Water causes swelling of seed, and is necessary for digestion, translocation, and growth.

Oxygen. Seeds must respire to germinate and must have oxygen for aerobic respiration. Lack of oxygen causes growth of anaerobic bacteria which cause rotting.

Temperature. Most seeds will not germinate if the temperature falls close to freezing or rises above 115°F. Most favorable temperatures are 70–85°F.

Food supply. Some small seeds (orchids) germinate only if an external food supply is available in the environment. In nature, fungi provide this nourishment.

Other external factors. Light, soil acidity, carbon dioxide, etc. also influence seed germination.

INTERNAL CONDITIONS.

Auxins. The presence of *auxins* (growth regulators) influences germination.

FOOD. Stored food must be sufficient.

COMPLETION OF DORMANCY. Dormancy is a period of relative rest required by most seeds before they can germinate. Dormancy may be due to: undeveloped embryos; thick seed coats which render absorption of oxygen or water difficult or which resist swelling and growth of embryo; necessity of "after-ripening," or gradual chemical changes in embryo. Dormancy is a means of carrying seeds through a period unfavorable to active growth.

Since seeds have low water content, they are resistant to many environmental factors which would be injurious to actively growing tissues.

Seed Viability. Most seeds remain viable, i.e., they possess the ability to sprout, for not more than five or six years. Some remain viable for only a few weeks (orchid). Others may retain viability for three or four hundred years (Indian lotus). Dry, cool storage conditions favor prolonged viability. Loss of viability seems to be due mainly to the coagulation of protoplasm.

SEED AND FRUIT DISPERSAL

Dispersal is the spread of seeds or fruits. Dispersal through various means is brought about by the following parts or types of seeds or fruits:

WINGS. Dispersed by wind. Elm, maple fruits; catalpa seeds.

PLUMES. Dispersed by wind. Dandelion fruits, milkweed seeds.

SPINES AND BARBS. Dispersed by animals and man, to whose fur or clothing they stick. Cocklebur fruits.

AIR SPACES OR CORKY FLOATS. Dispersed by water. Coconut.

MINUTE SEEDS. Blown by wind. Orchids.

FLESHY FRUITS. Eaten by animals, seeds scattered with feces.

NUTS. Buried in ground by squirrels, etc.

EXPLOSIVE FRUITS burst and scatter seeds. Touch-me-not, oxalis.

Chapter 17

Heredity and Variation

The ability to reproduce is one of the most striking characteristics of organisms, and is unique to the living world even though a self-reproducing machine is readily imaginable. The obvious fact that "like begets like" must be modified by the fact that generations of sexually reproduced organisms are not *exactly* alike, but are similar only within limits. There are variations among individuals.

ENVIRONMENT AND HEREDITY

Variations among individuals may be the result of environmental forces. Light, temperature, chemical composition of the environment, etc., may affect the form and behavior of an organism, but modifications that environmental forces impose upon an organism do not usually have any effect upon its heritable characteristics. Experiments made to find what effects environment can have on genetic material have been constantly negative. Some forces, however, can act directly upon the genetic material, but such forces (ionizing radiations, chemical mutagens, and possibly others) do not work through a visibly affected part. For example, a plant may have a non-green leaf because the leaf has been kept dark, but the offspring of the plant will have normal leaves. If reproductive cells of the plant are treated with X-rays, however, some change in the DNA may occur, with a resultant change in chlorophyll-synthesizing ability. To some extent, then, environmental forces may affect an organism's genetic quality, but the difference between a direct effect upon an organ and an effect upon DNA is enormous. No machinery exists for transfering information from an affected leaf, for instance, to pollen grains or eggs of the same plant.

SEXUAL AND NON-SEXUAL REPRODUCTION

Non-Sexual Reproduction. The simplest, and presumably oldest, kind of reproduction is non-sexual (asexual). Division (*fission*) of one-celled organisms, and liberation of single cells or groups of cells from a parent plant are types of non-sexual reproduction. No recombination of genetic characters is involved, and an organism produced in this way is potentially exactly like its parent; any structural or functional differences it may exhibit are imposed by its environment.

Sexual Reproduction. Sexual reproduction has one feature that non-sexual reproduction lacks: the meeting of two portions of DNA, with possible consequent recombination of physical characteristics in the resulting organism. Sexual reproduction may have such complex attendant features as the formation of flowers with eggs, pollen, seeds, and fruits; or it may be merely the fusion of two nuclei in a cell of a fungus, or the passage of a chromosome (or even a piece of a chromosome) from one bacterium to another. The meeting of two DNA components is the essence of sexuality.

MENDELIAN GENETICS

The first important advance in understanding the principles of inheritance came with the work of Gregor Mendel. Although his major contribution was published in 1866, it was essentially ignored until 1900, when it was rediscovered almost simultaneously by Correns, De Vries, and von Tschermak. Mendel's innovations were: (1) his attention to one pair of characteristics at a time, or a few pairs at most, rather than the whole mosaic of characteristics which make up a plant, and (2) his method of counting organisms, thus applying a statistical procedure for the first time to problems of inheritance.

A Monohybrid Cross. When Mendel chose to pay attention to one characteristic, such as seed color, he made a *monohybrid cross*. Clearly, when two plants are cross-bred, many characteristics may be mated, but when a breeder concerns himself with only one, the procedure is called a monohybrid cross.

LAW OF DOMINANCE.   In his work with garden peas, Mendel chose seven readily observable features, including size, color, method of flowering, and texture. When he fertilized a dwarf plant with pollen from a tall one, he found that all the resulting seeds in the F_1, or first filial, generation produced tall plants. These results caused him to formulate the *Law of Dominance:* when the possibility of two contrasting characteristics (e.g., tallness or dwarfness) exists in an organism, one characteristic (tallness) may be expressed to the exclusion of the other. The hidden characteristic is called *recessive,* and the expressed one is called *dominant.* Such a pair of alternate possibilities is a pair of *alleles.*

An organism possessing two identical alleles, as for tallness or for dwarfness, is *homozygous* for that character. If it possesses different alleles, as one for tallness and one for dwarfness, it is *heterozygous.* By convention, dominant traits are symbolized by capital letters (T for tallness), while the recessive allele is symbolized by the same letter, but in lower case (t for dwarfness). The observable condition of the organism is known as its *phenotype,* while the genetic condition is its *genotype.* Thus a pea plant which is phenotypically tall may be genotypically either homozygous (TT) or heterozygous (Tt). If the plant is phenotypically a dwarf, it can only be homozygous recessive (tt).

LAW OF SEGREGATION.   In a Mendelian monohybrid cross, a tall pea plant (TT), pollinated by a dwarf one, yields all tall plants (Tt) because the tall plant can produce only eggs with the factor T, the dwarf plant can produce only sperms with the factor t, and the resulting zygote must have both T and t. If a tall plant, heterozygous for tallness, is self-pollinated, it can yield three genotypes, because it can produce T-type and t-type eggs in equal numbers, and T-type and t-type sperms, also in equal number. The chances for a T-type egg to be fertilized by T or by t sperms are equal, and the same is true of a t-type egg. Consequently, one fourth of the zygotes will be TT, one fourth Tt, one fourth tT, and one fourth tt. Because of the dominance of tallness, all the resulting plants except the tt one will be phenotypically tall. This ratio of three phenotypically dominant to one recessive is typical of a monohybrid cross, and gave Mendel the basis for his *Law of Segregation.* This law states that hereditary determiners may come together in one generation and then segregate when that generation produces offspring.

For ease of handling the symbols, a "checkerboard," or matrix of squares, was devised by the geneticist Punnett. In this method, all possible sperm types are listed across the top of the matrix, the possible egg types down the left side, and the combinations in the squares. The monohybrid cross involving tallness, as shown in a Punnett's square, is symbolized thus:

	T	t
T	TT	Tt
t	tT	tt

If an organism is phenotypically dominant for a given character, its genotype cannot be determined by inspection, but it can be determined by crossing the organism of unknown genotype with a recessive. This is a *back cross* or a *test cross*. If a tall plant is homozygous, a test cross will produce all dominants, but if it is heterozygous, it will produce half dominants and half recessives.

A Dihybrid Cross. Mendel also considered two pairs of alleles simultaneously, and found that the segregation of one pair

	TY	tY	Ty	ty
TY	TYTY tall yellow	TYtY tall yellow	TYTy tall yellow	TYty tall yellow
tY	tYTY tall yellow	tYtY dwarf yellow	tYTy tall yellow	tYty dwarf yellow
Ty	TyTY tall yellow	TytY tall yellow	TyTy tall green	Tyty tall green
ty	tyTY tall yellow	tytY dwarf yellow	tyTy tall green	tyty dwarf green

of alleles (as tallness and dwarfness) was independent of the segregation of a different pair of alleles (as, for example, yellow color or green color of seeds). If Y represents a gene for yellow, and y represents green, then a tall, yellow-seeded plant, heterozygous for both characteristics, would have a genotype TtYy, and it could make four kinds of gametes: TY, tY, Ty, and ty. If such a plant is self-pollinated, it will show all possible combinations of phenotypes in the ratios shown in the Punnett's square on the previous page: 9/16 TY, 3/16 tY, 3/16 Ty, and 1/16 ty.

If a tall, yellow pea plant of unknown genotype is subjected to a test cross (dwarf, green), the phenotypic results might come out ¼ TY, ¼ tY, ¼ Ty, and ¼ ty. This would show that the plant must have been heterozygous for both pairs of alleles, as can be shown by a simple Punnett's square.

	ty
TY	TY ty
tY	tY ty
Ty	Ty ty
ty	ty ty

LAW OF INDEPENDENT ASSORTMENT. Such activity of inheritance factors led to the Mendelian *Law of Independent Assortment:* the manner of segregation of one pair of alleles is not determined by the segregation of other pairs of alleles.

Later discoveries necessitated modifications of the Law of Independent Assortment, the main addition being the qualification that the segregation of one pair of alleles on *one chromosome pair* is not affected by the segregation of pairs of alleles *on other chromosome pairs*.

GENES AND CHROMOSOMES

Linkage and Crossovers. An important exception to the Law of Independent Assortment was discovered when some characters, which were expected to segregate independently, showed non-Mendelian numbers of offspring. If an organism with genotype AaBb were test-crossed, the results might not yield the anticipated ratio of phenotypes (25% AB, 25% Ab, 25%aB, and 25% ab) but would give a high proportion of parental phenotypes and a low proportion of recombinations. For example, a result might be 45% AB, 5%aB, 5%Ab, 45% ab.

LINKAGE. Such discoveries suggested that some genes are linked together. Continued experimentation showed that organisms had whole groups of characters inheritable in linked groups. Lists of such linkage groups were prepared for a number of organisms, and the *number of linkage groups* equaled the *haploid number of chromosomes* in the species investigated.

Since linked genes do not remain permanently linked, but do show recombination in a definite percentage of crosses, it was assumed that genes are carried in linear sequence on chromosomes, and that chromosomes can exchange parts. Each gene is thought to occupy a specific place, or *locus* (plural, loci), on a chromosome. Thus, a chromosome carrying linked genes A and B may exchange its B part with a homologous chromosome carrying a and b. After the exchange, the first chromosome would carry A and b, and the second would carry a and B.

CROSSOVERS. An exchange which separates previously linked genes is a *crossover*. If the frequency of crossing-over between two linked genes is proportional to the distance between them, then by determining the percent of recombination between a number of *loci*, the distance between the genes and the order of their arrangement can be determined.

The distances are measured in *map units*, one map unit being equated with 1% crossover between linked genes. In the example described above, since 5% Ab and 5% aB recombinants were recorded, a total of 10% crossovers indicated that the locus for gene A (or a) is ten map units from that for gene B (or b). Linkage maps have been prepared for a few organisms, especially tomatoes, corn (maize), fruit flies, mice, and some fungi, bacteria, and viruses.

Chromosomes as Carriers of Genetic Factors. Linkage phenomena suggested that genes are specifically carried by chromosomes rather than by other cell components. Additional evidence came from the similarity of behavior between genetic factors (Mendelian segregation) and that of chromosomes (meiosis and fertilization). Still stronger evidence came with the discovery that microsopically visible chromosome morphology is associated with the sex of the individual. First noted in the plant, *Melandrium,* sex chromosomes were demonstrated in a number of animals. In some, one pair of unmatched chromosomes could be found in males only (*Drosophila,* mammals) or in females only (moths). Sex chromosomes are rare in plants because plants do not commonly have distinctly separate sexes.

MUTATIONS

From his observations on evening primroses, De Vries first described mutations as sudden, heritable changes in an organism. His "mutations" were actually morphological changes in chromosomes. Most mutations, however, are chemical changes in DNA, cannot be detected microscopically, and are recognizable only from their effects on some phenotypic character.

Chromosome Aberrations. Chromosome aberrations, sometimes called "chromosome mutations," are deviations from normal chromosome morphology. These include (1) *shifts,* in which a piece of a chromosome is displaced to a new position in the same chromosome; (2) *translocations,* in which a piece of a chromosome is transferred to a different chromosome; (3) *deletions,* in which a piece of a chromosome is lost; (4) *duplications,* in which a piece of chromosome is represented more than once; and (5) *inversions,* in which a piece of a chromosome is turned through 180 degrees in the same chromosome, reversing the sequence of loci in the inverted piece.

Gene Mutations. Unless otherwise specified, "mutation" means a change in a gene. The slightest possible change would be a change in a single pair of bases in a DNA molecule. Such a change, that is, a change in a single *muton,* would result in a change in the RNA which is synthesized at the affected place. The altered base in the RNA would mean that the triplet of RNA bases (i.e., the *codon*) would be associated with a different amino acid from the old, normal situation (Chap. 13). Thus, a protein at the ribosomal site would come to contain one new kind of amino acid. If the change made an appreciable difference in the behavior of the protein (e.g., if it were a critical enzyme), the organism containing the new protein would be phenotypically different from the pre-mutation organism.

Most mutations that have been found and studied are recessive, and are likely to be deleterious. In a large population, many deleterious, recessive alleles may remain in the *gene pool,* being expressed rarely because they rarely meet in a homozygote.

Although the word *gene* is still commonly used, no entirely acceptable definition of a gene has been devised. An increasingly detailed knowledge of the action of nucleic acids has shown that the concept of a gene as a discrete entity is too simple. Some possible approaches to clarification of the idea of a gene are (1) the functional approach, in which a gene is considered as the ultimate unit of cellular activity, or of mutability, and (2) the distributional approach, in which a gene is considered as the ultimate unit of recombination. However, the ability of any unit to act or to recombine is affected not only by its own DNA constitution but also by the position it occupies in relation to the rest of the DNA in a chromosome. Consequently, any attempt to define a gene as a discrete entity is incomplete and misleading.

BIOCHEMICAL GENETICS

Currently, fruitful investigations have been made through observing metabolic characteristics instead of morphological features especially in fungi, bacteria, and viruses.

The pink bread mold, *Neurospora,* reproduces sexually, and consequently can be used to produce hybrids, show recombinations, and segregation. It has an advantage over higher plants and animals in that reduction division results in haploid spores which germinate directly into haploid plants. Thus no heterozygous condition, and no masked recessives, are present to complicate genetic analyses. By artificially producing a number of mutants (using X-rays or ultraviolet light), investigators can demonstrate the presence or absence of specific enzymes, and the absence of an enzyme can then be shown to be heritable in the same fashion as a structural feature. This information led not only to the idea that a gene is essentially an enzyme-producer (the "one gene-one enzyme" hypothesis), but also to the elucidation of a number of detailed biochemical pathways.

The colon bacillus, *Escherichia coli*, has also been used for biochemical studies. Its advantage over higher organisms lies in the fact that it can be handled statistically in astronomic numbers. Thus extremely rare recombinations, which would be too infrequent to have a probability of appearing in only a few thousand offspring, can be detected and analyzed.

Viruses which attack bacteria are called *bacteriophages*. Like bacteria, they can be analyzed genetically, although the individual phage particles are too small to be seen with a conventional microscope. The most detailed separation of genetic factors presently available has been made with a series of phages (the T viruses) which attack *E. coli* and are detectable by the speed and manner of attack. Millions of virus particles can be considered in a single experiment.

Chapter 18

The Classification of Plants

The objects of plant classification are to arrange plants in groups for identification and to indicate, wherever possible, relationships among plants. The exact total number of kinds (*species*) of plants on the earth is not known; about 350,000 species are known at present. The science of classification is *taxonomy*.

The Basis of Classification. Until after the Renaissance, plants were classified chiefly on vegetative characters, such as growth habits, leaf structure, human utility, etc. At present, reproductive structures and behavior are the chief bases of classification. Emphasis is placed on reproductive features instead of on vegetative ones because these are less susceptible to the influence of environmental factors and are consequently more stable.

Systems of Classification. A *system* of classification is a complete arrangement of the major groups of plants or of all plants into a unified scheme. Different botanists have proposed different systems of classification, because complete knowledge of true relationships among certain plants is lacking and consequently opinions as to such relationships vary.

A *natural* system attempts to classify organisms on the basis of their evolutionary relationships. A natural system is the goal of taxonomy. As more facts about relationship are discovered, artificial systems are replaced by natural systems.

CLASSIFICATION UNITS

In the following listing, groups progress from more specific to broader categories.

Species. The species is the basic unit in the classification of organisms. A species is a *kind* of organism—e.g., a dog, a white

145

oak, a sugar maple, etc. A species may be defined technically as the *usually* smallest unit in the classification system; it is a group of individuals of the same ancestry, of similar structure and behavior, and of stability in nature; that is, the members of a species retain their characteristic features through many generations under natural conditions. One of the most dependable criteria for the delimitation of a species is interfertility among its individuals, although even this criterion is not entirely usable. In many groups of plants, species are sharply delimited, in other groups, there are no sharp lines between species because of intermediate forms. Sometimes there are different types (*varieties*) of organisms within a species; for example, the various kinds of dogs are all varieties of the dog species; such varieties do not maintain themselves in nature, but are maintained artificially by man.

Genus (pl. *genera*). A genus is a collection of closely related species. For example, the oak genus is made up of a number of species—red oak, white oak, pin oak, etc. All have certain common major characteristics, but differ in minor ways.

Family. A family is a group of closely related genera. For example, the oak family is made up of the oak genus (*Quercus*), the chestnut genus (*Castanea*), etc. The names of families end usually in "aceae," less frequently "ae"; e.g., the oak family is the *Fagaceae*. These genera have certain traits in common which show them to be related but there are also marked differences among them.

Order. An order is a group of closely related families which have certain common traits but which differ in certain respects. The names of orders end in "ales." The oak order is the *Fagales*.

Class. A class is a group of related orders. The names of classes end in "ae."

Division. A division is a group of related classes. The names of divisions end in "phyta."

CLASSIFICATION OF THE PLANT KINGDOM

Several systems of classification are in use, and systems change as new knowledge and new ideas develop. A system currently favored by many botanists, especially in the United States, is the following:

Subkingdom *Thallophyta* (plants not forming embryos)
Division 1—*Cyanophyta* (blue-green algae)
Division 2—*Euglenophyta* (euglenoids)
Division 3—*Chlorophyta* (green algae)
Division 4—*Chrysophyta* (yellow-green algae, golden-brown algae, diatoms)
Division 5—*Pyrrophyta* (crytomonads, dinoflagellates)
Division 6—*Phaeophyta* (brown algae)
Division 7—*Rhodophyta* (red algae)
Division 8—*Schizomycophyta* (bacteria)
Division 9—*Myxomycophyta* (slime molds)
Division 10—*Eumycophyta* (true fungi)

Subkingdom *Embryophyta* (plants forming embryos)
Division 11—*Bryophyta* or *Atracheata* (plants lacking vascular tissues)
  Class *Musci*—mosses
  Class *Hepaticae*—liverworts
  Class *Anthoceratae*—hornworts
Division 12—*Tracheophyta* or *Tracheata* (plants with vascular tissues)
  Subdivision *Psilopsida*—psilopsids
  Subdivision *Lycopsida*—club mosses
  Subdivision *Sphenopsida*—horsetails and relatives
  Subdivision *Pteropsida*—ferns and seed plants
    Class *Filicineae*—ferns
    Class *Gymnospermae*—cone-bearing plants and relatives
    Class *Angiospermae*—true flowering plants
      Subclass *Monocotyledonae*
      Subclass *Dicotyledonae*

Divisions 1 to 7 are called "algae" for convenience, but this is not the official name of a group, as in older systems. Similarly, divisions 8, 9, and 10 are sometimes called "fungi." Unlike fungi, the algae contain chlorophyll and manufacture their own food.

Scientific Names. Scientific names of plants are Latin in form and are regulated by agreements passed by botanists at International Botanical Congresses. The advantage of using Latin is that the elements of this language are known by most scientists and it constitutes an international language of science.

Scientific names are often descriptive of traits or places, commemorate people, are based on classical mythology, etc.

Each species of plant or animal has a scientific name composed of two words; the first, which is capitalized, is the name of the genus, the second, not capitalized, is the name of the species. This system of naming organisms is called the "binomial system." It was invented in the seventeenth century and was first used consistently on a large scale by Linnaeus. Following each scientific name is an initial or abbreviation which indicates the man who named that species. Examples of scientific names are *Zea mays* L. (corn), *Rhus cotinoides* Nutt. (smoke tree).

In studying the plant groups, students should organize their knowledge about each group according to this outline:

1. STRUCTURE.   General body structure.

2. REPRODUCTION.   Types and nature of reproduction; alternation of generations.

3. HABITATS AND DISTRIBUTION.   Environments commonly inhabited; occurrence in different parts of world.

4. IMPORTANCE TO MAN.   Importance in agriculture, industry, etc.

5. EVOLUTIONARY IMPORTANCE.   Relationships with other groups; relative position in plant kingdom.

6. COMMON REPRESENTATIVES.   Names of typical members of the group.

7. NUTRITION.   Plants may produce their own food, i.e., be *autotrophic*, using energy from light (photosynthetic types), or from chemical reactions (chemosynthetic types). Some plants must obtain preformed food, i.e., are *heterotrophic*. If they use food from a living host, they are *parasitic*, if from dead organic matter, they are *saprophytic*. Plants that live on other plants without parasitizing the substrate are *epiphytic*.

Chapter 19

Thallophyta: Algae

Thallophyta in general are characterized more by negative than by positive features, and are described in terms of the features they lack in comparison with the next Subkingdom, the Embryophyta. A scheme of classification which depends upon listing absent structures will obviously lead to some unnatural groupings. Botanists consider the Thallophyta a *polyphyletic* group, that is, a group of individuals, arranged in various *taxa* (categories used in classification) which have evolved from diverse ancestors so long ago that now their relationships are obscure or entirely lost. The *taxon* Thallophyta is still retained as a practical entity even though some subgroups have little demonstrable affinities with others. Nevertheless, some characteristics are:

1. Lack the characteristically complex organs: roots, stems, leaves.

2. Produce sex cells in simple, usually one-celled structures.

3. Have relatively undifferentiated tissues, without such highly specialized vegetative cell-types as xylem, fibers, cambium, sieve cells, epidermis, etc.

4. Do not produce young, immature individuals (*embryos*) inside the parent organisms.

5. Are with few exceptions microscopic, or at best relatively small, although some marine algae and spore-bearing bodies of fungi are macroscopic.

6. Are commonly aquatic and autotrophic, or terrestrial and heterotrophic, although many exceptions exist.

THE ALGAE

Algae consist of seven divisions, with some of these subdivided

149

by some authors. These include about 20,000 known species, all of which are autotrophic. Although they contain chlorophyll, many plants in these phyla are brown, red, or of some other hue as a result of the presence of pigments which mask the green color of chlorophyll.

Division Cyanophyta (blue-green algae).

STRUCTURE. Contain chlorophyll and a blue pigment (phycocyanin), both contained in microscopic granules, the *chromatophores*. A few species are red. No plastids. No nuclei. One-celled. Cells often held together in colonies by a mucilaginous secretion. Colonies are chiefly threadlike (filamentous), or are clusters or sheets of cells.

REPRODUCTION. Entirely asexual, chiefly by fission. A few species form resting "spores."

HABITATS AND DISTRIBUTION. Both in salt and fresh water, chiefly latter. On wet rocks, damp soil, etc. Some in vicinity of hot springs. Widely distributed.

IMPORTANCE TO MAN. Contaminate water supplies, furnish fish food, add organic matter to soil and increase soil fertility. Some poisonous to fish and domesticated animals. Some combine atmospheric nitrogen into organic molecules.

COMMON REPRESENTATIVES. *Oscillatoria, Nostoc.*

Division Euglenophyta (euglenoids).

STRUCTURE. Are one-celled plants which swim by means of flagella. Are spherical, ovoid, pear-shaped, etc. Each cell contains a nucleus and usually one or more chloroplasts. Green, brownish, reddish, or yellowish green in color. A red *eyespot*, a light-sensitive structure, is usually present. In many species, a primitive *gullet* is present, by means of which solid food may be ingested. Cells often remain together in colonies. Possess a mixture of plant and animal characteristics; non-green flagellates are often classified as animals.

REPRODUCTION. By cell division or by simple gametes.

HABITATS AND DISTRIBUTION. Are water plants, growing in oceans, lakes, ponds, etc. Widely distributed on earth's surface.

IMPORTANCE TO MAN. Food for fish and other water animals; pollute water supplies because of bad odors and taste which they give to water.

Evolutionary Importance.   Many biologists believe that flagellates are similar to first living organisms on the earth, that possibly both plants and animals have evolved from flagellate ancestors.

Common Representative.   *Euglena.*

Division Chlorophyta (green algae).

Structure.   Green in color. Chloroplasts and nuclei present. Some one-celled, some colonial, some multicellular. Most species are filamentous; some form sheets or clusters of cells, others are strictly unicellular. Some species have special *holdfasts,* which anchor the filaments on stones, etc.

Reproduction.   Both asexual (zoospores, fission, fragmentation) and sexual methods occur (isogamy if mating sex cells appear identical, *heterogamy* if they differ in size, content, motility, etc.).

Habitats and Distribution.   Grow in both salt and fresh water, on wet soil, fence posts, moist stones, tree bark, etc. Chiefly fresh-water. Widely distributed.

Importance to Man.   Pollute water, add oxygen to water and thus benefit fish, furnish food for fish, furnish human food, form carbonates and thus build rock deposits, used for theoretical studies in genetics and photosynthesis.

Common Representatives.   Some variations in structure and reproduction are shown by the following genera: (1) *Protococcus.* One-celled, found commonly on tree bark. Reproduction by cell-division only (Fig. 19/1). (2) *Ulothrix.* Filamentous, with holdfasts at bases of filaments. Grows in running water. Reproduces asexually by means of 4-ciliate zoospores, which are formed 2, 4, 8, or 16 per cell; each zoospore swims, attaches itself to a solid object in water, then grows into a filament. Reproduces also sexually by formation of 32 or 64 2-ciliate isogametes, which are similar to zoospores structurally, though smaller; gametes fuse in pairs and form zygotes. Each zygote, after a resting period,

Fig. 19/1.
Protococcus.

germinates, forms four zoospores, which swim, then attach themselves to solid objects and form filaments (Fig. 19/2). (3)

Spirogyra. Filaments growing in fresh water without holdfasts. One or more spiral chloroplasts per cell. Reproduces asexually by fragmentation; does not form asexual spores. Reproduces sexually by isogamy, as follows: two filaments, lying side by side, form protuberances from opposite cells; these protuberances push the threads apart, meet in pairs, fuse at tips into tubes; protoplasm (single gamete) of one cell moves through tube into opposite cell, fuses with protoplasm (gamete) of latter to form a thick-walled zygote, which after a resting period undergoes meiosis and grows into a new filament, which is haploid. Sometimes protoplasm of one cell fuses with that of an adjoining cell of same filament (Fig. 19/3). (4) *Oedogonium.* Filamentous, attached to solid objects in fresh water. Considerable differentiation among cells. Reproduces asexually by many-ciliate zoospores, which swim about, then attach themselves and grow into new filaments. Repro-

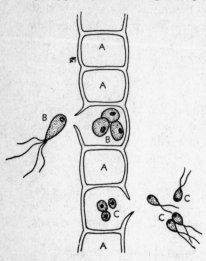

Fig. 19/2.   Ulothrix. *A. Cells of filament. B. Zoospores. C. Isogametes.*

Fig. 19/3.   Spirogyra. *A. Vegetative filament; a. spiral chloroplasts. B. Conjugating filaments; b. conjugation tubes, c. gamete moving through conjugation tube, d. zygote.*

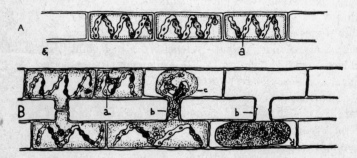

duces sexually by heterogamy; certain cells of filaments become *oogonia,* each of which contains a single egg; other cells become *antheridia,* each with two ciliated sperms; a sperm swims through a pore in the oogonium, fertilizes the egg in the oogonium. The thick-walled zygote rests for a time, then forms four zoospores, each of which, after liberation, grows into a new filament. The filaments and gametes are haploid, the zygote is diploid; meiosis occurs in the formation of zoospores which are haploid. In some species antheridia and oogonia develop in the same filament, in others, they are formed on separate filaments (Fig. 19/4).

EVOLUTIONARY IMPORTANCE OF GREEN ALGAE.   Some reproduce only by fission, others by zoospores, isogamy, heterogamy, etc. Gametes seem to have developed from zoospores, which they often resemble very closely in structure, though smaller in size (e.g., in *Ulothrix*); in *Oedogonium,* sperms are like zoospores structurally but smaller, eggs also resemble zoospores. Most primitive type of sexual reproduction involves union of isogametes like zoospores (*Ulothrix*); more advanced type involves union of heterogametes, which differ markedly in size (*Oedogonium*). In some algae, very slight size difference among gametes; transitional from isogamy to heterogamy. Beginnings of cellular differentiation in such vegetative cells as holdfasts.

Division Chrysophyta (yellow-green, golden-brown algae, diatoms).

STRUCTURE.   Mostly unicellular and colonial; few multi-cellular. Cell walls usually composed of two overlapping halves and impregnated with silica, thus having a harsh texture. Flagella

Fig. 19/4.   Oedogonium. A. Cells of filament. B. Oogonium with egg. C. Antheridium with sperms. D. Sperm. E. Fertilization pore.

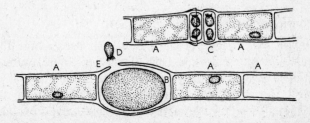

present or absent. Color of chlorophyll partly or wholly masked by yellowish or brownish pigments. Nuclei and plastids present.

REPRODUCTION. Asexual by cell division, zoospores, or other types of spores. Sexual reproduction, when it occurs, is isogamous.

HABITATS AND DISTRIBUTION. In damp soil and in fresh and salt water. Occur in enormous numbers in water, especially ocean water.

IMPORTANCE TO MAN. Most important are the diatoms, which have beautifully sculptured walls. Diatoms are used in filters in industry, polishes for metals, insulation for pipes and furnaces, dynamite, insulation in refrigerators, and sometimes in tooth-pastes. Much petroleum is of diatom origin. Diatomaceous earth is a rocklike deposit of dead diatom shells and is mined in several parts of the world to obtain material for the uses mentioned above. Diatoms constitute an important food source for fish and other aquatic animals.

COMMON REPRESENTATIVES. Most familiar are the diatoms.

EVOLUTIONARY IMPORTANCE. Exact relationships not known. Some botanists believe them to be related to flagellated, unicellular green algae.

Division Pyrrophyta (cryptomonads, dinoflagellates).

STRUCTURE. Microscopic, unicellular, nucleate, with plastids containing chlorophyll and yellow-brown pigment. Mostly flagellate, with one trailing flagellum and one encircling flagellum.

REPRODUCTION. Mostly asexual, with sexual method (conjugation) not definitely established.

HABITATS. Mostly marine, worldwide.

IMPORTANCE TO MAN. Of little direct importance, but of enormous indirect importance as primary food producers of the sea.

Division Phaeophyta (brown algae).

STRUCTURE. Largest of algae, most complex structurally, with considerable differentiation (some have sieve tubes), often reach lengths of several hundred feet. Have large holdfasts. Brownish pigment (*fucoxanthin*) present in addition to chlorophyll. Some filamentous. Many have stemlike, rootlike, and leaflike parts, often with air bladders which give buoyancy.

REPRODUCTION. Some reproduce asexually by zoospores. Sexual reproduction in some by isogamy, some by heterogamy. Alternation of generations in some species.

HABITATS AND DISTRIBUTION. Chiefly in cooler ocean waters, a few in fresh water. Grow attached to rocks, etc., mainly in shallow water.

IMPORTANCE TO MAN. Food for fish and other water animals, source of iodine, mineral salts used as fertilizers and in soap manufacture, food for man and cattle, used in medicinal preparations.

COMMON REPRESENTATIVES. (1). *Ectocarpus.* Filamentous, attached. Forms two types of zoospores— *diploid* and *haploid.* Former grow into diploid plants like their parents, forming more zoospores; latter grow into haploid plants which form isogametes and thus reproduce sexually. This is a primitive type of alternation of generation (see Chap. 12). Zygotes formed by haploid plants develop into diploid plants which form zoospores. Zoospores and gametes similar structurally. In some species, gametes alike in size, in others size differences occur. Thus, there are some heterogamous species.

2. *Fucus* (Bladder wrack). On rocks in intertidal zones of temperate seas. No asexual reproduction except by fragmentation. Mature plant is dichotomously branched thallus, with enlarged branch tips, often with air bladders. Thalli have holdfasts, stem-like parts. In enlarged tips there are small sunken cavities (*conceptacles*), tips of which appear as tiny blisters on branch tips. Sexual reproduction heterogamous. Conceptacles contain antheridia and oogonia; in some species, these are on separate plants, in others, are produced in same conceptacle. Antheridia produce up to 64 sperms, oogonia usually 8 eggs each. Sperms and eggs escape at maturity through pores at the tips of the conceptacles into sea water. A sperm fuses with an egg in water to form a zygote, which grows into a new plant. Plants are diploid; meiosis occurs at gamete-formation, as in animals, but it is a rare occurrence in higher plants.

3. *Kelps.* Large, massive, brown algae, found chiefly on Pacific Coast of U.S. Have holdfasts, often large air bladders.

*Fig. 19/5.
Diatom.*

Division Rhodophyta (red algae).

STRUCTURE. Chiefly feathery, branched-filamentous, sometimes ribbon-like with gelatinous surfaces. Smaller than most brown algae, rarely more than three or four feet long. Red in color, due to *phycoerythrin*, a red pigment present with chlorophyll.

REPRODUCTION. Asexually by non-motile spores, sexually by non-motile heterogametes. Male gametes are spherical, non-motile, produced in antheridia. Male gametes carried by water to female sex organs (*carpogonia*). Each carpogonium has an elongated tip (*trichogyne*), to which a sperm may adhere. Wall of trichogyne near sperm dissolves, nucleus of sperm passes down trichogyne and fuses with female gamete nucleus at base of trichogyne. Zygote produces numerous short branches, on which are borne *carpospores*. In some red algae, carpospores grow into new plants which bear carpogonia and antheridia. In others, carpospores grow into plants which look like plants that produce carpogonia and antheridia, but which form not gametes, but asexual spores in clusters of four (*tetraspores*). Tetraspores grow into plants which produce carpogonia and antheridia, with female and male gametes respectively. Life cycle in such red algae consists of two alternating phases: haploid, gamete-bearing plants; and diploid, tetrasporic plants. Meiosis occurs at tetraspore formation, so that tetraspores are haploid. Thus, alternation of generations in some red algae.

HABTATS AND DISTRIBUTION. Chiefly warmer waters of oceans. Few fresh water. Often at great depths.

IMPORTANCE TO MAN. Food for sea animals and man, used in shoe polish, cosmetics, glue, confections, jellies. Agar is a gelatinous derivative of certain red algae; used as culture medium, as laxative, in clarification in liquors.

COMMON REPRESENTATIVES. *Nemalion, Polysiphonia, Chondrus* (Irish "moss").

EVOLUTIONARY IMPORTANCE OF RED ALGAE. The chief evolutionary importance of red algae lies in their highly specialized reproductive methods and in the definite alternation of generations which occur in many of them.

SUMMARY OF ALGAE

1. Algae are ancient, simple green plants.
2. Algae lack roots, stems, leaves.
3. Most algae are aquatic.
4. Blue-green algae are simplest structurally, brown algae are most complex.
5. The algae are a heterogeneous group of plants, the exact relationships of which are not known.
6. Include unicellular, colonial, and multicellular plants.
7. In lower algae, reproduction is entirely asexual. In others, isogamy occurs, and in more advanced types, heterogamy.
8. Heterogametes probably evolved from isogametes.
9. Alternation of generations occurs in some algae, but is not generally well-established.
10. The green algae may be ancestors of higher plants. The brown and red algae are specialized terminal groups. The flagellates are possible descendants of the first living organisms.
11. The sex organs of algae are structurally simple.

Chapter 20

Thallophyta: Fungi

The plants known popularly as fungi constitute a mixed, poly-phyletic group with uncertain relationships. One system of classi-fication categorizes them in three divisions: the *Schizomycophyta* (bacteria), the *Myxomycophyta* (slime molds), and the *Eumyco-phyta* ("true" fungi). All these share the common characteristics of thallophytes. The origins of fungi are unknown, but they apparently either evolved from some widely different ancestors, or arose so long ago that intermediates have become extinct and their relationships cannot be accurately determined. None of the present types gave rise to higher forms.

THE FUNGI

Division Schizomycophyta (bacteria).

STRUCTURE. Unicellular, often forming sheetlike or filamen-tous colonies. Smallest cellular organisms; average size is about 2 x ½ micron; some even smaller. Three shapes: spheres (*cocci*), rods (*bacilli*), and spirals (*spirilla*). Rods and spirals often have flagella. Cytoplasm simple structurally; primitive nuclei present, simpler than those of higher plants. Have thin walls, often with surrounding envelopes (Fig. 20/1). Mitochondria lacking, res-piratory activities carried on by membranes. Photosynthetic species have chlorophyll in minute chromatophores. Some species obtain energy for food manufacture by oxidation of inorganic substrates (chemosynthesis).

REPRODUCTION. By *fission,* which may occur as often as once every 20 minutes. If a bacterial colony maintained its most rapid rate of fission for 24 hours, about 4,700,000,000,000,000,000,000 bacteria would result from the original parent cell. This maximum

rate is not maintained indefi-
nitely because of lack of sufficient food and accumulation of
toxic wastes. Sexual reproduction
occurs in some.

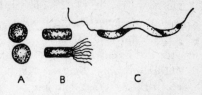

Fig. 20/1. Bacteria. A. Spheres.
B. Rods. C. Spiral.

SPORE FORMATION. Certain
kinds of bacteria form resting
spores, usually one spore per cell.
Spores are very resistant to un-
favorable external factors which might kill ordinary active cells.
Some biologists do not regard this as reproduction because usually a bacterium produces only one spore; thus, it usually does
not increase numbers of bacteria but is a method of carrying bacteria through an unfavorable period. When conditions become
favorable, spores germinate, each growing into an active bacterium.

HABITATS AND DISTRIBUTION. Most widely distributed of all
organisms: in air, soil, water, on and in other living organisms.
Chief requirements for their growth are water, favorable temperatures, organic matter, and presence of oxygen (*aerobic*
species) or absence of oxygen (*anaerobic* species).

IMPORTANCE TO MAN. Some activities of bacteria are harmful
to man, others beneficial. Beneficial effects are much more important.

Harmful Effects. (1) Cause diseases of man—typhoid, cholera,
pneumonia, lockjaw, tuberculosis, etc. (2) Cause diseases of domesticated animals—hog cholera, anthrax, tuberculosis, etc. (3)
Cause diseases of plants—fire-blight of pears and apples, citrus
canker, wilt diseases of tomatoes, potatoes, melons, cucumbers,
etc. (4) Cause spoilage of food. (5) Decompose materials: fabrics,
petroleum products, etc.

Beneficial Effects. (1) Industrial uses—manufacture of vinegar, cheese, alcohols, sauerkraut, silage, etc. Curing of vanilla,
tea, coffee, cocoa, tobacco; tanning of leather, retting of flax, etc.
(2) Decompose dead bodies and waste products of plants and
animals. Organic compounds in these are broken down by bacteria
and other fungi into simpler substances—CO_2, salts, etc.—which
can be used again by green plants in food synthesis. Clear earth
of dead bodies. (3) Maintain soil fertility, especially with refer-

ence to nitrogen. (4) Produce antibiotics. (5) Theoretical investigations in genetics, biochemistry, biophysics.

BACTERIA AND NITROGEN OF THE SOIL. Several types of bacteria are important in nitrogen transformations in the soil:

Ammonifying Bacteria. Decompose proteins into ammonia, which then forms ammonium compounds in the soil.

Nitrifying Bacteria. Convert ammonium compounds into other nitrogen salts. There are two kinds of nitrifying bacteria: *nitrite* bacteria (*Nitrosomonas*) which convert ammonium compounds into *nitrites* (NO_2 compounds); and *nitrate* bacteria (*Nitrobacter*) which convert nitrites into *nitrates* (NO_3 compounds). *Nitrosomonas* and *Nitrobacter* make their own food from water and CO_2, using energy obtained by oxidizing ammonium salts and nitrites (*chemosynthesis*). Of all nitrogen compounds, nitrates are most important in nutrition of green plants.

Nitrogen-Fixing Bacteria. Take nitrogen gas (which most green plants cannot use) from the air and build it into proteins. There are two types of these bacteria: those which live free in the soil (*Azotobacter* and *Clostridium*), and those (*Rhizobium*) which live in the root tissues of green plants, chiefly legumes. The latter type usually forms swellings, known as nodules. Plants with such N-fixing bacteria on their roots slow down N-loss of soils in which they grow.

Denitrifying Bacteria. Convert nitrogen salts in the soil into gaseous nitrogen which escapes into the air. This process, especially common in poorly drained soils, causes a loss of soil fertility.

Division Myxomycophyta (slime molds). A diverse group of organisms of uncertain affinities, ranging in size from microscopic to a few inches across. They are probably not all related even to one another, and are as animal-like at some times as they are plant-like at other times. They are probably related to protozoa.

STRUCTURE. Slime molds may be either *cellular* or *plasmodial*. In cellular slime molds, amoeba-like cells make contact with one another to form an aggregation of discrete cells all working together as an individual. The cells are sometimes known as "social amoebae." In plasmodial slime molds, one or more amoeba-like cells flow together to make a multinucleate, protoplasmic body which flows with internal currents, plus general mass movement, over moist surfaces. In the motile stages, the organisms feed like true amoebae by engulfing food, usually bacteria.

REPRODUCTION. The ways in which cellular and plasmodial slime molds produce spores are basically different, but the result is the formation of fungus-like bodies which bear spores. The spores germinate into amoebae, or in some species into motile cells with flagella.

HABITATS AND DISTRIBUTION. Worldwide in tropical and temperate climates. On decaying vegetation, in soil, and in living plants.

IMPORTANCE TO MAN. Except for a few plant parasites (e.g., club-root of cabbage), they are economically unimportant. The cellular slime molds are useful in theoretical research on cellular differentiation and biological control.

Division Eumycophyta (true fungi). The bodies of most members of this phylum consist of fungus filaments, or *hyphae*, which are long, slender, cottony structures. A mass of hyphae is a *mycelium*. The bodies of most species are multicellular, and most species possess some type of sexual reproduction, in addition to asexual reproduction by spores, budding, or some other asexual process. The cells of true fungi contain nuclei, a feature which distinguishes them readily from bacteria. The true fungi are usually separated into four classes: Phycomycetes, Ascomycetes, Basidiomycetes, and Deuteromycetes.

Class Phycomycetes (alga-like fungi).

STRUCTURE. Resemble certain green algae in structure. Bodies composed usually of hyphae which lack cross walls and thus have the form of continuous tubes. These hyphae are multinucleate. Hyphae are not organized into bodies of definite form but usually grow as irregular, cottony masses upon substratum.

REPRODUCTION. Reproduce asexually by motile zoospores and by non-motile spores. Sexual reproduction in some species is isogamous, in others, heterogamous.

HABITATS AND DISTRIBUTION. Some species are parasites and grow in the tissues of their host plants (*Albugo*, or white rust, on horseradish, and *Phytophthora*, or late blight, on potatoes). Many species are saprophytic and aquatic; some of these, such as water

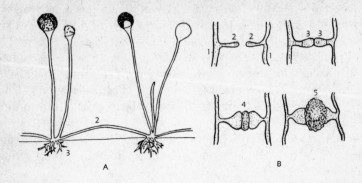

Fig. 20/2. Rhizopus (*black bread mold*). *A. Portion of* Rhizopus *mycellium; 1, Sporangiophore with asexual sporangium and spores; 2, Stolon; 3, Rhizoids. B. Stages in sexual reproduction of* Rhizopus: *1, Hyphae; 2, Progametes; 3, Isogametes; 4, Fertilization; 5, Zygote.*

molds (*Saprolegnia* and other genera) live on sticks, dead leaves, etc., in water or in soil. Some species are parasites on fish. Other species grow on fruits in transit, on bread, and on other food-stuffs, whose decay they cause.

IMPORTANCE TO MAN.   (1) Cause diseases of several important plants (downy mildew of grapes, white rust of mustards, potato blight, brown rot of lemons, etc.). (2) Cause spoilage of food, diseases of fish, decay of dead organic matter. (3) Produce industrially important chemicals.

EVOLUTIONARY IMPORTANCE.   Thought to be closely related to certain green algae, from which they have possibly developed.

COMMON REPRESENTATIVES.   A common species is *Rhizopus nigricans* (black bread mold), which forms a mass of hyphae on stale bread, fruits, and other foodstuffs (Fig. 20/2). Some of the hyphae (*stolons*) grow horizontally over the substratum; these send other hyphae (*rhizoids*) into the substratum; other hyphae (*sporangiophores*) grow upright. The latter develop enlarged sporangia at their tips. Each sporangium bears numerous asexual spores, which are released by the rupture of the sporangial wall and which grow into new hyphae. *Rhizopus* also reproduces sexually by an isogamous process similar to that of *Spirogyra*, as follows: two hyphae in contact with each other form lateral protuberances which press against each other. A wall is formed across each hypha a short distance behind the tip. The wall between these two cells breaks down and the two protoplasmic masses fuse. The zygote grows in size, and after a dormant period grows into a short sporangiophore, with an asexual sporangium at its tip. The spores released from the sporangium grow directly into new hyphae. In *Rhizopus nigricans,* certain hyphae produce one kind of gamete, others produce the second kind, a condition known as *heterothallism.* Only when both kinds of physiologically distinct hyphae are present does zygote formation occur. There are also *homothallic* algal fungi, in which both types of gametes are produced on the same hyphae.

Other common genera are: *Mucor* (black mold); *Saprolegnia* (water mold); *Plasmopara* (downy mildew); and *Phytophthora* (potato blight fungus).

Class Ascomycetes (sac fungi).   Largest class of fungi, with about 40,000 species.

STRUCTURE.   Most species have hyphae with cross walls, usually with one nucleus per cell. A few species are unicellular (yeasts). In some species, the hyphae have been organized into definite, often fleshy, bodies; in others, hyphae form cottony growths of indefinite extent.

REPRODUCTION.   Asexually by budding *conidiospores,* and by fragmentation. The most characteristic reproductive structures are sacs, or *asci* (sing., *ascus*), which are formed at the ends of certain specialized hyphae (Fig. 20/3). Each ascus usually produces eight *ascospores,* which grow directly into new hyphae after their release from the ascus. Asci develop differently in different species. In *Pyronema,* antheridia and oogonia are present; an oogonium with many nuclei develops at the end of a hypha and forms a trichogyne; a multinucleate antheridium forms on another hypha, comes in contact with trichogyne; intervening walls break down, protoplasm of antheridium moves into oogonium and fuses with protoplasm of latter; male and female nuclei pair off, each pair moving into a hypha, of which a number grow from oogonium; paired nuclei move into tip of hypha, are separated from rest of hypha by a wall; the two nuclei fuse, following which the fusion-nucleus divides

Fig. 20/3. Ascus with ascospores.

three times, forming eight nuclei; the terminal cell becomes an ascus, the eight nuclei are surrounded by bits of cytoplasm and become ascospores. In some species, an oogonium may be present, but no antheridium; in others, there are no differentiated sex organs.

HABITATS AND DISTRIBUTION. Some are parasites, growing in the tissues of certain host plants. Some are saprophytes, growing in soil on decaying organic matter, spoiling fruits and other food-stuffs, etc. They are widely distributed on the earth.

IMPORTANCE TO MAN. (1) Cause many familiar diseases of economic plants—peach leaf curl, Dutch elm disease, chestnut blight, apple scab, etc. (2) Industrial uses—manufacture of alco-holic beverages, cheese, "raising" of bread dough, etc. (3) Food spoilage. (4) Human food—truffles, morel. (5) Decay of dead organisms and their wastes. (6) Commercial production of organic chemicals. (7) Production of antibiotics.

COMMON REPRESENTATIVES. The major groups of sac fungi are:

Yeasts—one-celled, non-filamentous. Reproduce chiefly by bud-ding, less frequently by ascospores. Important in brewing, bak-ing. Secrete a number of enzymes which convert glucose to alco-hol and CO_2 (Fig. 20/4).

Cup Fungi—hyphae organized into fleshy, cup-shaped asco-carps, inside of which asci are formed. Chiefly saprophytes in rich soil, decaying wood, etc.

Powdery Mildews—parasites, chiefly on leaves of green plants (lilac, clover, etc.). Form whitish patches of hyphae on leaves, with tiny black *perithecia*, which contain asci.

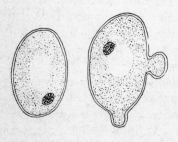

Fig. 20/4. Yeast cells, showing budding.

Blue and Green Molds—sapro-phytes on old leather, jellies, spoiling fruit, potatoes, etc. Hy-phae form indefinite growths, producing conidiospores and as-cospores in large numbers. *Peni-cillium* is common blue mold, *As-pergillus* produces blue or black spores. Some *Penicillia* are im-portant in cheese manufacture and production of penicillin.

Truffles—have fleshy, edible ascocarps.

Morels—cone-shaped, mushroom-like ascocarps.

EVOLUTIONARY IMPORTANCE OF SAC FUNGI. Some similarity exists between red algae and certain sac fungi with respect to reproductive structures and activities. Some botanists believe there is a relationship between these groups. Others believe them to be descendants of Phycomycetes.

Class Basidiomycetes (basidium fungi).

STRUCTURE. Have hyphae with cross walls. In lower forms, hyphae form indefinite growths, in higher forms, hyphae organized into definite, often fleshy bodies. Cells frequently contain two nuclei which divide simultaneously during growth.

REPRODUCTON. Asexual reproduction may be by *conidia*, cells constricted at ends of hyphae and carried away by wind, water, or animals. Characteristic sexual reproduction is by *basidiospores*. In a terminal binucleate cell, the *basidium*, nuclear fusion is followed immediately by meiosis. The four resultant haploid nuclei pass out through slender tubes, the *sterigmata*, into the spores (Fig. 20/5). Genetic recombinations and segregations are accomplished in the basidia. Basidia may be one-celled or variously partitioned by walls.

HABITATS AND DISTRIBUTION. Many species are parasitic and grow within the tissues of their hosts. The saprophytic species are found in rich soil, on decaying logs, etc. The basidium fungi are widely distributed on the earth.

IMPORTANCE TO MAN. (1) Edible species—some mushrooms and puffballs. (2) Cause serious plant diseases—rusts, smuts, etc. (3) Cause wood rotting. (4) Cause decay of dead organisms and the wastes of other organisms.

COMMON REPRESENTATIVES. The chief groups of basidium fungi are:

Smut Fungi. Parasites on cereal grains, chiefly infecting flowers and forming large numbers of black *chlamydospores* in the grains, which are often much enlarged and distorted. These spores usually rest till

Fig. 20/5. Basidium with basidiospores. A. Basidium. B. Basidiospores. C. Hypha.

following spring, then produce short hyphae, each of which bears usually four basidiospores, which reach host plants and infect them, either in seedling stage or later in flowers. Basidiospores form mycelium in host. Smuts do much damage to oats, corn, wheat, and other cereals. Formaldehyde treatment of seeds kills smut spores.

Rust Fungi. Parasites which cause serious diseases of wheat, oats, rye, and other cereals, pine, apples, etc. Called rusts because of reddish spores formed on surface of diseased tissues. Some rust fungi live on a single host, others require two different host species to complete their life cycle; the former species are called *autoecious*, the latter *heteroecious*. A common heteroecious rust is the wheat rust fungus (*Puccinia graminis*) the life cycle of which follows :

Hyphae in diseased wheat plants form, during the summer, blisters of orange, one-celled summer spores (*uredospores*) which are carried by wind to other wheat plants, which they infect. Later in the season, the hyphae form, in dark-colored blisters, winter spores (*teliospores*), which are two-celled, thick-walled, resting spores that remain on stubble and straw over winter. In the following spring, each cell of teliospore forms a short hypha which produces four basidiospores, which are carried by wind to leaves of the common barberry (not the cultivated barberry). The basidiospores form hyphae in the barberry leaves; these hyphae in seven to ten days form flask-shaped *spermogonia*, each of which contains hyphae with small cells (*spermatia*) at their tips. These are exuded onto the surface of the leaf by an opening in the spermagonium. There are two kinds of spermatia: *plus* spermatia formed from the hyphae produced by *plus* basidiospores, and *minus* spermatia formed by hyphae produced by *minus* basidiospores. When a spermatium of one kind meets a hypha of the other kind, fertilization occurs, resulting in the development of hyphae which produce cup-shaped *aecia*, which open on the lower side of the leaves. The aecia contain *aeciospores* which are carried by wind to young wheat plants, which they infect. Thus, the life cycle is completed.

Other heteroecious rusts are: cedar-apple rust (on cedars, apples, hawthornes, etc.), white pine blister rust (on white pine,

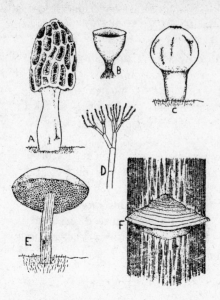

*Fig. 20/6. Some representatives of higher fungi. A. Morel. B. Cup
fungus. C. Puff Ball. D. Spore-bearing hyphae of* Penicillium. *E.
Stalked pore fungus. F. Bracket type of pore fungus on tree trunk.
(A, B, and D are Ascomycetes; C, E, and F are Basidiomycetes.)*

gooseberry, and currant). Control measures include elimination
of alternate hosts from the vicinity of the desired plants, and by
breeding of rust-resistant varieties.

Gill Fungi—mushrooms. Mycelium grows usually saprophytically underground or in decaying wood, and periodically form
fleshy sporophores (*mushrooms*) of characteristic size, form, and
color. A mushroom consists of a stalk and an umbrella-shaped
cap, on underside of which are radiating *gills*, which bear basidia
and numerous basidiospores. Mushrooms are classified on basis of
spore color; there is no infallible rule for distinguishing edible
from poisonous species.

Pore Fungi—often similar to mushrooms, but underside of cap
has pores within which basidia are formed. Common wood-
rotting fungi, chiefly saprophytic. Some edible.

Tooth Fungi—bear basidia on fine, toothlike masses of hyphae.

Puffballs—spherical, pear-shaped, etc. Basidiospores borne internally, inside a tough surface covering. Covering ruptures or has a pore for escape of spores. Most puffballs are edible when young. Mostly saprophytic.

Class Deuteromycetes (fungi imperfecti). This is a large, heterogeneous group of organisms of mixed relationships. They do not constitute a natural biological assemblage, but are classed together, for convenience, on the basis of the fact that in these organisms sexual reproduction is not known. Some have been demonstrated to be the non-sexual phase of ascomycete or a few basidiomycete genera. They are, except for bacteria, the most common terrestrial organisms, and may be found on plants, on and in animals, in air, in soil, and in water. They are prime agents of decay, and cause innumerable diseases of wild and cultivated plants. Imperfect fungi can cause diseases of lungs, ears, skin, and nerves of animals. The best known fungus diseases are athlete's foot and ringworm.

LICHENS

Lichens are associations of certain algae (blue-greens and greens) with fungi (chiefly sac fungi) in a state of *symbiosis* (mutual benefit). The fungi obtain food from algal cells, absorb and retain water, some of which algae use in photosynthesis. The algae and fungi usually reproduce simultaneously, forming bodies composed of algal and fungus cells, which are capable of growing into new lichens. Lichens are common on rocks, tree bark, fence posts, etc. There are three main types of lichens:

FOLIOSE—flat, leafy or thallus lichens.
CRUSTOSE—thin, hard crusts, especially common on rocks.
FRUTICOSE—erect, branched growths.

SUMMARY OF FUNGI

1. Fungi are thallus plants; have no roots, stems, or leaves.
2. Fungi lack chlorophyll, and are chiefly parasites and saprophytes. A few species are chemosynthetic.
3. Cells of higher groups of fungi organized into hyphae.
4. In lower fungi, reproduction chiefly asexual. In higher groups, both asexual and sexual reproduction occur. Both isogamy and

heterogamy occur. The chief reproductive structures are spores, which are scattered by wind, water, insects, and birds.

5. Lichens are associations of algae and fungi living symbiotically.

Chapter 21

Embryophyta: Division Bryophyta

The subkingdom *Embryophyta*, which comprises plants that form embryos, are separated into two divisions: *Bryophyta* (mosses and their relatives), which lack the vascular tissues xylem and phloem; and *Tracheophyta*, which possess these tissues.

GENERAL FEATURES OF BRYOPHYTA

Bryophyta are relatively simple green land plants. They are small and inconspicuous, with some differentiation of tissues, but no definite xylem, phloem, cambium, etc. Some are thalloid, others have leaflike and stemlike portions. They have *rhizoids* which anchor plants and absorb materials from soil; rhizoids thus have the functions of roots, but differ from roots in that they are less complex structures. All Bryophyta have definite alternation of generations; the gametophyte is the larger, more conspicuous generation, the sporophyte is smaller, simpler, and partly or wholly parasitic on the gametophyte. The sex organs are *antheridia* (produce sperms) and *archegonia* (produce eggs), which are more elaborate than the relatively simple organs of Thallophyta. Fertilization occurs *inside* the archegonium and the zygote develops in the archegonium; thus, egg, zygote, and embryo sporophyte receive protection and nourishment from archegonium and surrounding gametophyte tissue. Sperms reach the eggs in the pear-shaped archegonia by swimming through water. The Bryophyta are limited chiefly to shaded, moist habitats. There are about 23,000 species of Bryophyta.

Class Musci (mosses).

Structure.   Mosses usually grow vertically, as compared with

170

the commonly horizontal growth of liverworts. Mosses have a distinct stemlike structure, with small, green, leaflike appendages; these are not considered true stems and leaves because they lack the characteristic vascular tissues of the stems and leaves of higher plants. The "leaves" of mosses have midribs which distinguish them from those of leafy liverworts. From the base of the stem portion grow rhizoids which anchor the moss plant and absorb water and minerals. Moss plants rarely exceed six inches in height. The moss plant as seen with the naked eye is the major part of the gametophyte generation (Fig. 21/1).

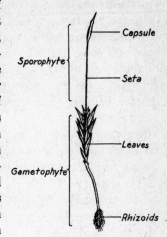

*Fig. 21/1.   Moss.**

REPRODUCTION. Archegonia and antheridia are borne at the tips of the leafy shoots, in some species at the same tip, in others on different plants. There are usually sterile hairs among and around the sex organs. The ovoid antheridia bear numerous sperms; each elongated, flask-shaped archegonium has a single egg. When mature antheridia are wet, the sperms emerge and move to the archegonia by swimming through water. Sperms swim down the canal of an archegonium, and one sperm fertilizes the egg, which then by growth develops into the young sporophyte within the archegonium. The *foot* of the sporophyte becomes embedded in the apex of the leafy shoot; the stalk (*seta*) of the sporophyte elongates; at the tip of the seta is the *capsule,* within which the spores are formed. On the capsule there is usually a lid (*operculum*). When the lid is removed, there is in most mosses a *peristome,* or ring of teeth which bend outward when they are dry, scattering the spores, and bend inward when they are moist, preventing spore dispersal. The *calyptra* is the remnant of the old archegonium which covers the upper part of the capsule. Meiosis occurs in the capsule when the spores are formed; the

---

* Reprinted by permission from *Outline of General Biology,* by Gordon Alexander, published by Barnes and Noble, Inc.

spores are thus haploid and constitute the first stage of the gametophyte generation; when the spores are exposed to the proper
environmental conditions, they germinate and form alga-like filaments of cells. Each filament is called a *protonema*. After a time,
a protonema develops buds, from which the leafy shoots grow.
These then produce archegonia and antheridia and the life cycle
is repeated. The spores, protonemata, and leafy shoots with their
sex organs and gametes constitute the gametophyte generation;
the zygote, foot, seta, and capsule constitute the sporophyte. As
in liverworts, the gametophyte generation is more conspicuous,
larger, and more complex structurally; as in liverworts, the sporophyte depends on the gametophyte for its nourishment, either
partly or wholly.

Asexual reproduction may occur in mosses by growth of new
shoots from old shoots.

Habitats and Distribution.   Mosses require water for fertilization and are poorly equipped for conserving water; thus, they
are limited to moist habitats. A few species grow in arid locations.
Mosses are world-wide in distribution. Many of them grow on
damp rocks, on the barks of trees, and in dense stands on the
soil, producing familiar carpets of vegetation.

Importance to Man.   Mosses are of little direct importance
to man. They are of some value in reducing soil erosion, and they
furnish food for certain kinds of animals. Peat-moss (*Sphagnum*)
"leaves" have many large cells with circular openings; as a result, peat-moss absorbs liquids greedily and holds them tenaciously; this property makes it valuable as an absorbing medium

*Fig. 21/2.   Liverwort thallus. A. Sporophyte. B. Rhizoids. C. Upper
surface. D. Lower surface with rhizoids.*

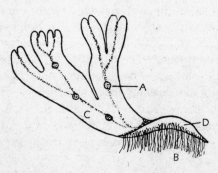

used for preserving cut flowers, etc.; ground peat-moss is used as a lawn dressing; compressed, dried peat is used as fuel in many parts of the world. Mosses are of some value in filling in lakes and in aiding in the formation of soil.

EVOLUTIONARY IMPORTANCE. Mosses show many evolutionary tendencies similar to those of liverworts, from which they differ chiefly in their upright growth and their greater structural complexity. The mosses are usually considered as a terminal evolutionary group, which has given rise to no other groups.

Class Hepaticae (liverworts).

STRUCTURE. In some species, the body is a flat, horizontal, branching, green thallus, with rhizoids growing from the lower surface into the soil. A liverwort thallus rarely exceeds three or four inches in length by one-half inch in width. Some species have partly erect bodies, with stemlike and leaflike parts. Cellular differentiation is slight in most liverworts (Fig. 21/2).

REPRODUCTION. A liverwort plant, as seen with the naked eye, is the major part of the gametophyte generation. On this plant are produced antheridia and archegonia; these may be partly embedded in the thallus or borne on stalks rising from the upper surface of the thallus. In some species, the archegonia and antheridia are borne on separate thalli. At maturity in the presence of liquid water, the antheridia open and discharge their many motile sperms. These swim to an archegonium in which, at maturity, certain cells in the neck disintegrate, leaving a canal which extends to the egg-cavity (*venter*) of the archegonium. An archegonium contains a single egg. Several sperms may enter an archegonium, but only one fertilizes the egg. The fertilized egg (zygote) begins to grow and develops in the archegonium into the sporophyte generation, which is attached to the gametophyte by a *foot* and which consists, in addition, of a short stalk, with a *capsule* at its apex; the capsule is a structure of varying shape, within which spores are formed (Fig. 21/3). Meiosis occurs in the formation of spores, so that the four spores produced from a spore mother cell are haploid. When the spores are mature, the capsule opens or disintegrates, freeing the spores, which fall to the ground and, under favorable conditions, grow into new thalli. The spores are the first structures of the gametophyte generation,

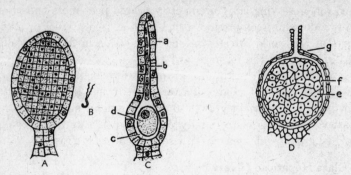

Fig. 21/3. Reproductive structures of liverworts. A. Antheridium.
B. Sperm. C. Archegonium; a. neck, b. neck canal, c. venter, d. egg.
D. Sporophyte; e. spores, f. sporophyte jacket, g. remnants of arche-
gonium.

which is haploid; the gametophyte generation ends with gamete-
formation (the gametes are haploid); the sporophyte, or diploid
generation, begins with the zygote and continues until the spores
are formed. The sporophyte is smaller, less complex than the
gametophyte and is dependent upon the latter for food.

In many liverworts there is vegetative reproduction by *gemmae*,
small bits of tissue which develop on the upper surface of the
thalli; when these gemmae become detached from the parent
thallus they grow directly into new thalli.

HABITATS AND DISTRIBUTION. Liverworts are poorly equipped
for conserving water. Thus they are limited to moist places,
usually away from direct sunlight. They live for the most part
on soil, though a few species are aquatic. They grow in many
places in the world.

IMPORTANCE TO MAN. Liverworts are of little importance to
man. They may break the force of raindrops and thus reduce soil
erosion. Some of them are colonizers of rocks, which other kinds
of plants cannot inhabit. Some are used by animals as food.

COMMON REPRESENTATIVES. A common genus is *Marchantia*,
in which the archegonia and antheridia are produced on separate
plants on vertical stalks which arise from the upper surfaces of
the thalli. An antheridial stalk bears a flattened, slightly concave
disc at its apex; the antheridia are embedded in the disc and open
at maturity onto the upper surface of the disc. An archegonial

stalk bears a number of umbrella-like ribs at its apex; the archegonia are borne on the undersides of these ribs, with their necks
pointing downward. Sperms reach the archegonia in running or
splashing water. The small sporophytes develop in the archegonia
and hang in a pendulous position, from which they drop their
spores at maturity.

Another common genus is *Riccia,* in which antheridia and
archegonia are borne on the same thallus; they are not borne on
stalks as in *Marchantia,* but are embedded in the thallus near its
midrib.

Evolutionary Importance of Liverworts. The liverworts
are the simplest true land plants of the present time. They resemble certain algae in their thalloid structure and are believed
to have evolved from algae. However, they have made certain
evolutionary advances over the algae: the development of multicellular sex-organs which protect the egg, zygote, and young
sporophyte, etc. Though they are land plants, the liverworts are
restricted to moist environments by their structural simplicity,
their lack of ability to conserve water, and their dependence upon
water for fertilization.

Class Anthoceratae (hornworts). This is a small group of
plants which superficially resemble liverworts. They differ from
liverworts and mosses in being structurally much more complex.

SUMMARY OF BRYOPHYTA

1. Bryophyta are the most primitive green land plants. Their
simple construction and dependence on water for fertilization
limit them to moist habitats.

2. Sex organs of Bryophyta are multicellular.

3. The sperms of Bryophyta are ciliated and swim to the
archegonia through liquid water.

4. Fertilization and zygote development occur within the
archegonium; thus, the zygote and young sporophyte are protected and nourished by the gametophyte.

5. Alternation of generations occurs in all Bryophyta. The
gametophyte is larger and more complex than the sporophyte. The
gametophyte makes its own food and also nourishes the sporophyte which cannot make sufficient food for its nutrition.

6. The gametophyte generation begins with spores and includes the main body of liverworts and mosses and their sex organs and gametes. The sporophyte generation begins with the zygote, ends with meiosis in the capsule during spore formation from spore mother cells. Spores are haploid and begin the gametophyte generation.

7. Bryophyta probably evolved from some algal ancestor. Bryophyta are considered as rather specialized, though primitive, land plants which have not evolved into other types of plants. Their lack of vascular tissues prevents their reaching a large size and growing in any except a very moist habitat.

8. The proportion of sterile (non-spore-producing) tissue of the sporophyte increases markedly from the most primitive liverworts through the mosses.

9. In lower Bryophyta, there are no special mechanisms for spore dispersal; in higher types, there is the hygroscopic peristome which promotes spore dissemination.

Chapter 22

Tracheophyta: Psilopsida, Lycopsida, and Sphenopsida

The division Tracheophyta includes all plants which contain the vascular tissues xylem and phloem. All these plants form embryos, and all have alternation of generations, in which the sporophyte is larger and structurally more complex than the gametophyte. Four subdivisions are included in the Tracheophyta: the Psilopsida, the Lycopsida, the Sphenopsida, and the Pteropsida.

PSILOPSIDA

The Psilopsida are known mainly from fossils. Only three species are known to survive as representatives of what was once a large and important group of plants.

Structure. The plant body is rather simple, having a dichotomously branched stem but lacking roots. Leaves are usually absent or are very small. Rhizoids perform the functions of roots. No cambium is present, and thus there is no secondary growth.

Reproduction. The plant just described is a sporophyte, bearing sporangia on the stems. The sporangia produce one type of spore, which germinates on the soil to form a small, greenish gametophyte. This bears archegonia and antheridia. Sperms reach the archegonia by swimming through water. Fertilization of the egg occurs inside the archegonia, and the zygote grows into an embryo sporophyte within the archegonium.

Habitats and Distribution. Extinct species grew in many parts of the world. The living species are tropical and subtropical land plants.

Importance to Man. Psilopsida have no economic value.

Common Representatives. Plants of the genus *Psilotum* are occasionally seen as curiosities in greenhouses.

Evolutionary Significance.   Although this subdivision is insignificant in the world's living flora, it is thought to be very important from the evolutionary standpoint. Certain features of the structure and geological history of this group indicate that it probably originated from green algae and that it is the source group from which such groups of higher plants as ferns, horsetails, club mosses, and seed plants were derived. Among the important fossil genera of this subphylum is *Rhynia;* from plants of this type the higher groups of plants may have evolved. (See Fig. 24/1 in Chapter 24.)

LYCOPSIDA

This subdivision includes about 900 living species, as well as many extinct species known only as fossils. The living species are usually called *club mosses.*

Structure.   Living plants of this subphylum are rarely more than three feet in height; many are elongated creepers on the soil. They have vascular tissues, true roots, stems, and small leaves which are most often spirally arranged on the stems. Branch gaps are present, but there are no leaf gaps. Some extinct species were large trees.

Reproduction.   A club moss plant is a sporophyte, which bears sporangia on the upper surfaces of leaves clustered usually at the tips of stems in cones or strobili. The cones may be as long as two or three inches. Some club mosses are *homosporous* (bearing one type of spore and one type of gametophyte), and others are *heterosporous* (having two types of spores and two types of gametophytes). Sporangia-bearing leaves are called sporophylls.

HOMOSPOROUS CLUB MOSS—*Lycopodium.*   Each sporophyll bears a sporangium with many spores, all alike. These spores fall to the ground and grow into small gametophytes on the soil. These are usually green, though in some species they are non-green and are saprophytes. The gametophytes bear archegonia and antheridia; sperms which are biciliate, as compared with the multiciliate sperms of horsetails and ferns, swim through water to the archegonia, within which fertilization occurs. The zygote develops into the young sporophyte, which has a temporary structure, the *suspensor,* which pushes the young sporophyte deeper into the tissue of the gametophyte, thus enabling the young sporophyte to

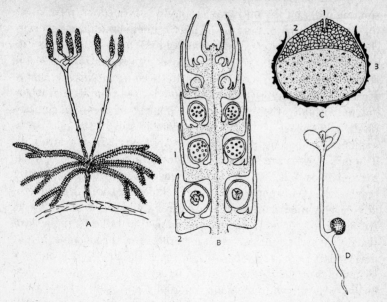

Fig. 22/1. Club mosses. A. Plant of Lycopodium, *with strobili. B. Longitudinal section of* Selaginella *strobilus; 1. microsporophyll with microsporangia and microspores, 2. megasporophyll with megasporangia and megaspores. C. Megagametophyte of* Selaginella; *1. archegonium with egg, 2. megagametophyte, 3. megaspore wall. D. Megagametophyte with young sporophyte* (Selaginella).

derive nourishment more advantageously from the gametophyte. The suspensor is a structure formed in seed plants and club mosses, but lacking from horsetails and ferns. The young sporophyte develops into the mature sporophyte which then produces spores.

HETEROSPOROUS CLUB MOSS—*Selaginella.* The sporophytes bear strobili of sporophylls at the branch tips. Each strobilus has two kinds of sporophylls, the lower of which are *megasporophylls,* the upper are *microsporophylls.* These bear, respectively, *megasporangia* (one per sporophyll) and *microsporangia* (one per sporophyll). A megasporangium produces four *megaspores,* a microsporangium produces usually 64 smaller *microspores.* These spores are formed as a result of meiotic divisions of spore mother cells and are thus haploid. A microspore begins to grow into a *microgametophyte* (male gametophyte) before it leaves its sporangium, and the microgametophyte is formed entirely within the spore. The major part of the microgametophyte forms sperms; the microgametophyte lacks chlorophyll and depends for its growth upon food stored in it while it was still attached to the sporophyte. A megaspore begins to form a *megagametophyte* (female gametophyte) before the spore is shed from its megasporangium; usually the complete development of the megagametophyte takes place while the megaspore is still in the megasporangium, which remains on its megasporophyll in the strobilus. The microspores are shed and fall down to the megasporophylls, reaching positions on or near the megasporangia. When water reaches the microspores with their included mature microgametophytes, the microspore wall is ruptured and sperms are released. An opening appears in the megasporangium wall and the sperms reach the archegonia which are borne near the opening of the megasporangium wall. Sperms can reach the archegonia only by swimming through water. A sperm fertilizes an egg and the zygote begins to develop into the young sporophyte.

The first division of the zygote forms two cells: the *suspensor* cell, which develops into the suspensor, and the *embryo* cell, which develops into the young (embryo) sporophyte deep in the tissue of the female gametophyte, from which the young sporophyte derives food. This development of a young sporophyte within the megaspore often occurs while the megaspore is still within the megasporangium. After a short time, the female gametophyte with its growing, attached sporophyte is shed and falls to the ground. The root of the young sporophyte enters the soil, leaves develop, and the sporophyte becomes independent of the megagametophyte. When the sporophyte reaches maturity, it then forms strobili, with two kinds of sporophylls, two kinds of sporangia, two kinds of spores, and two kinds of gametophytes.

The most significant features of reproduction in *Selaginella* are: heterospory, the formation of strobili bearing two kinds of sporophylls, reduction in the size of the mega- and microgametophytes, development of gametophytes *inside* the spores, retention of these spores with their gametophytes inside the sporangia, formation of a suspensor, dependence of the gametophytes for nourishment on the sporophyte (Fig. 22/1).

Habitats and Distribution.   Club mosses grow in many parts of the world, usually, though not always, in rather moist places.

Importance to Man.   Unimportant. Some ("ground pines") are used in Christmas decorations. Lycopodium powder used in fireworks and flashlight powders. Club mosses of Carboniferous helped form coal.

Common Representatives.   See Reproduction.

Evolutionary Significance.   Club mosses show many similarities with seed plants—heterospory, suspensor, retention of gametophytes within sporangia, etc. These indicate some relationship with true seed plants. Most botanists believe club mosses are a group parallel with seed plants but not so highly developed. Lycopsida apparently evolved from Psilopsida, reached their zenith in the Carboniferous Period, and are on the road to extinction.

SPHENOPSIDA

Like Psilopsida and Lycopsida, this subdivision reached its zenith in past geological ages and is doomed to extinction. Only 25 living species are known. Many extinct species have been discovered as fossils in rocks.

Structure.   Sphenopsids have vascular tissues, true roots, stems, and leaves which are usually small and, in living species, scalelike. The leaves are whorled. Stems, which are jointed and hollow, are green and carry on photosynthesis. Living species contain silica, which gives them a harsh, rough texture. Most living species are not over four feet tall, but some of the extinct species were large trees.

Reproduction.   A horsetail plant (*Equisetum*) is a sporophyte. Specialized spore-bearing leaves are borne in cone-shaped clusters (*strobili*) at the apices of the stems. Each sporophyll is hexagonal and bears five to ten saclike sporangia, within which the spores are produced. Reduction division occurs in the formation of spores from spore mother cells. Each spore has attached to it two *elaters,* ribbon-like structures which open and wind with changes in atmospheric humidity and thus cause spore-dissemination. The spores, which are all alike, fall to the ground and grow into *prothalli,* which are quite similar to those of ferns, except

Fig. 22/2. Horsetail (fertile shoot). A. Roots.
B. Stem. C. Scale leaves. D. Strobilus (cone).

that they are usually more slender and often branched. The prothalli bear archegonia and antheridia similar to those of ferns. These sex organs may be borne on the same prothallus or on separate prothalli. As in ferns, the sperms reach the eggs in the archegonia by swimming through water. A sperm fuses with an egg to form a zygote, which, attached to the gametophyte, grows into the sporophyte plant; the young sporophyte is nourished for a short time by the gametophyte until it develops sufficient green tissue to make its own food. At maturity, the sporophyte forms strobili, the sporophylls of which develop spores; thus the life cycle is completed. In some species, the strobili are borne only on certain specialized branches.

Habitats and Distribution.   The horsetails are found in many parts of the tropics and are also widely distributed in the temperate zones. They grow usually along river banks, around the margins of lakes, and frequently in ditches and along railroad embankments.

Importance to Man.   There is little economic use of horsetails. Their deposits of silica make them useful as a scouring material; the dried, powdered stems are used in certain kinds of scouring powders. Some of the horsetails of the Carboniferous period aided in coal formation.

Common Representatives.   There is only one genus (*Equisetum*) of living horsetails. In past geological ages, there was a much greater number of kinds of horsetails, which are now extinct. These ancient horsetails often reached much greater sizes than those attained by living horsetails.

Evolutionary Significance.   Sphenopsida apparently originated from Psilopsida, reached their zenith in the Carboniferous Age, and are on their way to extinction.

Chapter 23

Tracheophyta: Pteropsida

Pteropsida is the dominant subdivision of the earth's present vegetation. The plants vary from small herbs to great trees. Most are land plants, but some species are aquatic. All have vascular tissues and true roots, stems, and leaves. The leaves, which are larger than those of Lycopsida and Sphenopsida, apparently arose by flattening and transformation of branch systems. (Leaves of Lycopsida are outgrowths from surface of stem; those of Sphenopsida are probably transformed minor branches.) The sporangia of Pteropsida are borne on lower surfaces or margins of the leaves.

Pteropsida is separated into three classes: *Filicineae* (ferns); *Gymnospermae* (conifers and allies); and *Angiospermae* (true flowering plants). The latter two classes, which produce seeds, are often called seed plants.

CLASS FILICINEAE (FERNS)

Structure. A mature fern plant is the sporophyte generation. It consists of roots, stems, and leaves, similar functionally and structurally to those of seed plants. The stems of most temperate zone species are horizontal and subterranean and are thus rhizomes. In some tropical species (the tree ferns) the stems are erect and trunklike. The stems of ferns contain xylem and phloem, but usually no cambium; pericycle and endodermis surround each vascular bundle; sclerenchyma is located just beneath the epidermis and also in strands through the stems; parenchyma cells surround these other tissues, and epidermal cells form the surface layer and persist as the protective layer of the stem. The primary root lasts but a short time and dies when it is replaced by

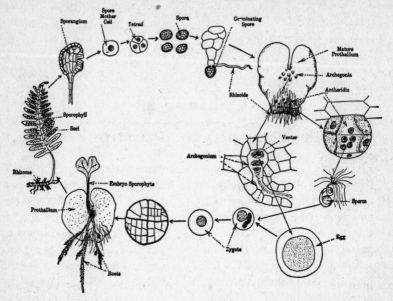

Fig. 23/1. Diagrams showing stages in the life cycle of a true fern (Polypodium).*

adventitious roots from the stem. The leaves (*fronds*) of most ferns are compound; the leaflets are known as *pinnae*, the petiole from which they arise is termed the *rachis;* in doubly compound leaves, the leaflets of the second order are called *pinnules*. The leaves of ferns have an internal structure similar to that of seed plant leaves; the veins, stomata, and guard cells are much like those of seed plants. The leaves of ferns are green and make food; in most species, the leaves also produce *sporangia* (Fig. 23/1).

Reproduction. The sporophyte reproduces by spores formed in *sporangia* which are borne on the green leaves, usually on the underside, or on special leaves which produce sporangia but do not make food. The sporangia are borne in clusters called *sori* (singular, *sorus*) which appear as brownish dots or streaks on the leaves. A typical sporangium consists of a short stalk, with a compressed, circular, biconvex capsule at its tip; the capsule

---

* Reprinted by permission from *Textbook of General Botany,* by R. M. Holman and W. W. Robbins, 3rd ed., published by John Wiley & Sons, Inc.

wall is one cell thick, with a row of thick-walled cells (the *annulus*) extending from the stalk about three-quarters of the way around the capsule. The inner and radial walls of the annulus cells are thickened. At the terminus of the annulus is a group of thin-walled *lip cells*. Within the sporangium are produced *spore mother* cells, usually 16, which by meiosis produce haploid spores, usually 64 in number. As the sporangium matures, it begins to dry out; the annulus, because of its thickened inner and radial walls begins to straighten and tears the lip cells, rupturing the capsule and scattering spores. This movement of the annulus is often sudden and spring-like; with changes in atmospheric humidity, the annulus may undergo more bendings after its first movement, thus scattering more spores.

After being shed, spores fall to the ground and under favorable conditions germinate, a single spore growing by repeated cell divisions into a *prothallus* (gametophyte). A prothallus is a thalloid structure, usually heart-shaped and rarely more than one-quarter inch in diameter, resembling the thallus of a liverwort. It is one cell thick on the margins, several cells thick in the center, and usually grows horizontally on damp soil. From the lower side of the prothallus grow rhizoids which anchor the prothallus to the soil and absorb water and minerals. In the thicker part of the prothallus are borne antheridia and archegonia, both of which open onto the lower surface of the prothallus. These sex organs are smaller and simpler in structure than those of *Bryophyta*.

An antheridium produces usually 32 sperms, which are set free by the rupturing of the antheridial wall; an archegonium has a short neck and a single egg in its base; when the egg is mature, the canal cells decompose, forming substances which attract the sperms. Sperms reach the archegonia only by swimming through water; several sperms may enter an archegonium, but only one fertilizes an egg. The union of a haploid sperm with a haploid egg forms a diploid zygote which develops by numerous cell divisions into the young sporophyte; the young sporophyte has a foot embedded in the gametophyte, by means of which it absorbs food from the gametophyte; the growing sporophyte soon produces a primary root, a primary leaf, and a small stem. The primary root and leaf disintegrate after a time, and the stem develops the adult leaves and adventitious roots of the mature sporophyte. The young sporophyte is nourished by the gametophyte until it develops enough leaf tissue to make its own food. The sporophyte generation includes all parts of the life cycle beginning with the zygote, to the meiotic divisions of spore mother cells in the sporangia; the gametophyte generation begins with spores and ends at fertilization.

Habitats and Distribution. Ferns grow chiefly in shaded, moist places, although there are some species which thrive in dry, exposed habitats. The greatest number of individuals and of fern species occurs in the moist regions of the tropics, where tree ferns reach heights of 30 or 40 feet. Some ferns are limited to definite ranges of soil acidity or alkalinity.

Importance to Man. Living ferns are of little importance to man, except as ornamental plants. Some species produce abundant, long, epidermal hairs which are used as a stuffing and packing material. The trunks of tree ferns are used for construction purposes in the tropics. Many ferns and their relatives of the Carboniferous period contributed largely to the formation of coal deposits.

Common Representatives. Among the common genera of ferns are: *Adiantum*—maiden-hair ferns. *Polypodium*—polypody fern. *Polystichum*—Christmas fern. *Camptosorus*—walking fern. *Nephrolepis*—Boston fern. *Water ferns* are heterosporous, in contrast to the previous homosporous genera; common water fern genera are *Marsilia* (water fern) and *Salvinia* (duckweed fern). *Botrychium*—rattlesnake fern.

Evolutionary Significance of Ferns. The ferns represent a line of evolution different from that of Bryophyta, in that their sporophyte is dominant and more complex and that they are better adjusted to a land habitat. The gametophytes and sex organs of ferns are simpler than those of Bryophyta. In some ferns there occurs *heterospory,* a condition found in all seed plants. The presence of heterospory, the development of complex xylem and phloem, and the retention of the young sporophyte within the gametophyte indicate a close relationship between ferns and seed plants. Most botanists believe that seed plants probably evolved from some group of extinct fernlike ancestors. It is believed that ferns evolved from ancient members of the Psilopsida. (See Fig. 24/1 in Chapter 24.)

CLASS GYMNOSPERMAE
(CONIFERS AND RELATIVES)

Structure. All species are woody plants, chiefly evergreen trees with needle- or scale-like leaves. Their tissues frequently contain resins and aromatic oils. Most species have a dominant main trunk, with much smaller branches.

Reproduction. Gymnosperms bear reproductive structures called *strobili*, or *cones*. A cone is a reproductive branch bearing usually non-green sporophylls. In most species, mega- and microsporophylls are borne on separate cones, called, respectively, seed cones and pollen cones; in most species, both types of cones are produced on the same plant. In a few species, both types of sporophylls are borne on the same cone. The details of reproduction in most gymnosperms are exemplified by the genus *Pinus* (pines), as follows (Fig. 23/2):

CONES. In pines, the seed cones are rather large and woody, with woody megasporophylls (*female cone scales*). Each megasporophyll bears two ovules on its upper surface. The pollen cones are much smaller than the seed cones and are not woody, with many small microsporophylls (*stamens*). Each stamen produces numerous pollen grains.

SPORANGIA AND SPORES. An ovule is a megasporangium surrounded by an integument; inside an ovule is a megaspore mother cell, which, by reduction division, forms four cells, one of which becomes a megaspore. Each microsporophyll bears two microsporangia, inside which microspore mother cells produce microspores.

GAMETOPHYTES. The megaspore inside an ovule undergoes many cell divisions to form the mega- (female) gametophyte; this contains many food storage cells and several archegonia near one end of the ovule, the end in which there is present a tiny pore (*micropyle*). These archegonia are much smaller than those of lower groups of plants; each contains a single egg. A microspore grows into a few-celled micro- (male) gametophyte (dispersed pollen grain); pollination occurs at this stage, the developing male gametophytes (pollen) being carried from the dehiscing microsporangia of the microsporophylls to the ovules on the megasporophylls by wind. The gametophytes lack chloro-

*Fig. 23/2. Cones of the western yellow pine. A. Staminate cone, natural size. B. Ovulate cone, one-half natural size. C. Single scale from latter, bearing two winged seeds.**

phyll and are thus dependent upon the sporophyte for their nourishment. The formation of the two kinds of cones, the development of the microspores and young microgametophytes and of the megaspores and young megagametophytes, and pollination occur during the first growing year of cone formation. During the second year, the archegonia are formed in the megagametophytes in the ovules, fertilization occurs, and the seeds mature.

FERTLIZATION. In pollination, the pollen grains are carried by wind directly to the surfaces of the ovules in the seed cone. When a pollen grain lands at the micropylar end of an ovule, it begins to form a tube during the first summer. Usually this tube is not completed during the first summer, but continues its growth during the next spring. Following pollination, the scales of the seed cone grow and are sealed together by pitch exuded from the cone; the developing ovules are thus held tightly inside the cone. The tube enters the micropyle, enters the nucellus (megasporangial wall) of the ovule, and discharges two sperms into an archegonium in the megagametophyte. One sperm fertilizes the egg, the other sperm and remaining cells of the pollen tube disintegrate. The fertilized egg (zygote) has the diploid chromosome number and develops into the embryo of the seed (matured ovule). The upper end of the embryo becomes a suspensor which pushes the embryo into the food-storage tissue of the female gametophyte, from

* Reprinted by permission from *Outline of General Biology*, by Gordon Alexander, published by Barnes and Noble, Inc.

which the embryo derives its early nourishment. Several zygotes may be formed in a single ovule, but as a rule only one embryo reaches maturity. Fertilization and seed development occur during the second year of cone growth; the scales of the seed cone separate at the end of the second year and shed their seeds, which germinate the third year in the soil.

SEED STRUCTURE.   A mature pine seed consists of the remainder of the nucellus, the seed coat, or matured integument, the endosperm, or food-storage tissue, which is a part of the female gametophyte, and the embryo, which consists of an epicotyl, a hypocotyl, and three to fifteen cotyledons. The germination of the seed is much as in flowering plants. (See Chapter 14.)

Habitats and Distribution.   Gymnosperms grow in many parts of the world. Some are found in the tropics, but their greatest development is in the cool parts of the temperate zones. They grow in a wide variety of habitats; some are desert plants, others grow only in regions of great humidity.

Importance to Man.   Economic products of gymnosperms are: lumber (soft woods), resins, turpentine, tars, aromatic oils, tannins, ornamental trees, edible seeds, etc. Great forests of pines, firs, etc., are important in preventing soil erosion.

Evolutionary Significance.   The living gymnosperms are the remnants of a formerly much larger and more varied group of plants. Many gymnosperms are known only in fossil form; such species are now extinct. Some of these extinct forms had leaves and stems like those of ferns, but with sporophylls and seeds similar to those of modern gymnosperms. Such species were intermediate between ferns and seed plants and are sometimes called "seed ferns." It is believed that the modern gymnosperms evolved from such seed-fern ancestors. The modern gymnosperms are not considered as ancestors of the angiosperms but are regarded usually as a more or less parallel evolutionary group.

Common Representatives.   Modern gymnosperms are separated into several orders, of which the more important are:

CONIFERALES, or conifers, the largest order of gymnosperms, including pines, spruces, firs, cedars, Douglas fir, redwoods, California big tree, bald cypress, yew, etc.

GINKGOALES, maiden-hair trees, belonging to the genus *Ginkgo*.

CYCADALES, palm-like plants, generally considered the most

primitive living seed plants. The sperms are ciliated, an unusual condition in seed plants. Cycads are dioecious.

CLASS ANGIOSPERMAE
(TRUE FLOWERING PLANTS)

Structure. The characteristic structural features of angiosperms have been described in Chapters 6 through 16. As in Gymnosperms, the visible plants are the sporophytes; the gametophytes are tiny, non-green, and usually visible to the naked eye only with difficulty.

Reproduction. Reproduction in angiosperms has been described in Chapters 15 and 16. There remains but to homologize the floral structures involved in reproduction in terms of alternation of generations. A flower, like a cone, is regarded as a specialized twig or branch, bearing reproductive leaves or sporophylls. The essential sporophylls are the stamens and carpels. Each stamen (microsporophyll) bears usually four pollen sacs (microsporangia), within which the spore mother cells and pollen grains (mature microspores) are produced. When a pollen grain carried by wind or insects reaches a stigma, it forms a pollen tube (microgametophyte), which grows down through stylar and ovary tissues until it reaches a megagametophyte (within an ovule). Each carpel (megasporophyll) encloses a number of ovules (megasporangia plus integuments), within which the spore mother cells and spores are produced. In each ovule, one megaspore develops into a megagametophyte (embryo sac). A microgametophyte sends two sperms into the megagametophyte, within which double fertilization occurs, as described in Chapter 15; the endosperm nucleus develops into the food-storage tissue of the seed, the zygote into the embryo (young sporophyte). As in gymnosperms, the gametophytes are non-chlorophyllous; they are very tiny, and depend nutritionally upon the sporophyte. Reduction division occurs in the formation of spores from spore mother cells and thus the gametophytes are haploid; the diploid chromosome number is restored by fertilization, the sporophyte being diploid. A striking difference between the megagametophytes of gymnosperms and of angiosperms is the lack of archegonia in the latter.

Habitats and Distribution. Angiosperms are the most widely distributed of all green plants on the land areas of the earth's surface. In addition, many species are aquatic.

Importance to Man. Angiosperms are the most important sources of foods, fibers, wood, rubber, drugs, and many other plant products useful to man. They constitute the chief part of the earth's vegetational cover and are most important as holding the soil, nourishing wild and domesticated animals, etc.

Common Representatives. The angiosperms are separated into two subclasses: Monocotyledonae and Dicotyledonae.

MONOCOTYLEDONAE. This subclass includes cattails, grasses, sedges, palms, lilies, irises, tulips, orchids, bananas, and many others. The distinguishing characters of monocotyledons are: one cotyledon per seed, flower parts in threes or multiples thereof, scattered vascular bundles in stems, usually no cambium, and mostly with parallel-veined leaves.

DICOTYLEDONAE. This includes willows, oaks, elms, maples, apples, roses, buttercups, petunias, phlox, violets, sunflowers, etc. Distinguishing characteristics: two cotyledons per seed, flower parts chiefly in fives and fours, rarely in twos or threes, vascular tissues in cylinders or regularly arranged bundles, cambium present, leaves mostly net-veined.

Evolutionary Significance. The angiosperms are the highest, most complex and apparently the most recently evolved major group of plants. They are best adapted of all groups of plants to a variety of environmental conditions. Some of them are very ancient, for angiosperm fossils (e.g., willow, sassafras, etc.) have been found in old rock strata. Most angiosperms are, however, apparently relatively recent plants. The woody angiosperms are generally considered to be older and possibly more primitive than herbaceous angiosperms. It is believed that angiosperms originated from some now extinct group of gymnosperms.

SUMMARY OF MAJOR DIFFERENCES BETWEEN ANGIOSPERMS AND GYMNOSPERMS

1. Gymnosperms bear seeds exposed on the surfaces of sporophylls. Angiosperms bear seeds enclosed by inrolled sporophylls (carpels). A matured carpel-base (ovary) of an angiosperm is termed a fruit.

2. Gymnosperms bear eggs in reduced archegonia in their female gametophytes. Angiosperms have eggs free in their female gametophytes, with no archegonia.

3. In the pollination of gymnosperms, pollen grains land directly upon the surfaces of ovules and thus the pollen tubes grow only a short distance. In angiosperms, pollen grains land upon a stigma, and pollen tubes must grow a considerable distance through style and ovary tissues before they reach the ovules, which are enclosed by the ovary.

4. All gymnosperms are wind-pollinated. Most angiosperms are insect-pollinated, some are wind-pollinated.

5. The endosperm of gymnosperm seeds is part of the female gametophyte and is thus haploid. The endosperm of angiosperm seeds develop from the fusion of a sperm with two embryo sac nuclei and is thus triploid.

6. Gymnosperms usually do not have sterile leaves associated with their sporophylls. Most angiosperms have sterile leaves (sepals and petals) associated with their sporophylls (stamens and carpels).

7. In most gymnosperms, sporophylls are on separate cones. In angiosperms, they are in the same cluster (flower).

8. Most angiosperms have vessels in their xylem, whereas such cells are lacking in most gymnosperms.

9. Gymnosperms are all woody perennials, whereas angiosperms include both woody and herbaceous (annual and biennial) species. Annuals grow from seed and form seed in a single year. Biennials require two years from seed planting to seed production. Perennials live for several to many years, forming seeds each year after reaching maturity.

SUMMARY OF FEATURES OF SEED PLANTS

1. The characteristic reproductive structures of seed plants are seeds. These are produced by cones or flowers. A cone or flower is a specialized reproductive twig.

2. Seed plants are not dependent upon water for fertilization. A pollen tube carries the sperms to the eggs.

3. The sporophyte generation is dominant and independent except for a brief time in its early development. The game-

tophyte generation is much reduced in size and structure, lacks chlorophyll, and is nourished by the sporophyte.

4. Seed plants are heterosporous, producing microspores and megaspores, which develop respectively in microsporangia and megasporangia which are borne on microsporophylls and megasporophylls. These spores grow respectively into microgametophytes and megagametophytes.

5. Megaspores are never released from their megasporangia, but develop into megametophytes within the megasporangial walls. The zygote and young sporophyte (embryo of the seed) likewise are retained within the megasporangia. A mature seed is a megasporangium, with its integuments, which enclose the female gametophyte (or its remnants), the megasporangial wall (nucellus), the young sporophyte (embryo), and endosperm.

6. Seed plants are the dominant plants on the earth.

Chapter 24

Evolution

Evolution is a process of change. Organic evolution is the historic process of change by means of which organisms have reached their present state. Organic evolution may also be defined as the changes which result in the development of new types of organisms or the extinction of other types.

There are three chief types of evolution:

1. PROGRESSIVE EVOLUTION, in which there is an increasing complexity in structure and differentiation.

2. RETROGRESSIVE EVOLUTION, in which there is a degeneration or decrease in complexity.

3. PARALLEL OR CONVERGENT EVOLUTION, in which different types of organisms show similar evolutionary changes, usually in similar sequence, often under similar environmental conditions.

EVIDENCES FOR ORGANIC EVOLUTION

Biologists agree generally that evolution has occurred and is still occurring, but often disagree concerning the explanations of data. Evidences regarded by biologists as proofs of the phenomenon are:

Comparative Morphology (Anatomy). Studies of structure show similar developments in different types of organisms, and similarities are believed to indicate relationships. E.g., strobili of club-mosses and gymnosperms, archegonia of Bryophytes, Pteridophytes, etc.

Comparative Embryology. This study shows similarities in origin and development in embryos of various plant and animal groups, e.g., similarities in development of young sporophytes of club-mosses, pine, angiosperms. In connection with embryology, the Biogenetic Law should be mentioned. It states that the development of an individual recapitulates the development of the race; e.g., the gametophyte of a fern is a thallus-like structure similar to the thallus of a liverwort or to certain green algae; the cycads have ciliated sperms similar to those of Pteridophytes, etc. The Biogenetic Law is not exact or dependable, and its validity is doubted by many biologists.

Geological Records.   Fossilized remains show that different kinds of plants and animals have lived at different periods in the world's history and that there lived in past ages plants and animals which are now extinct. Many living plants show striking resemblances to some of these fossil species.

Comparative Physiology.   This study shows that certain physiological processes and products are very similar in certain types of organisms, e.g., photosynthesis in green plants, resins in conifers, glycogen in fungi, aromatic oils in mint family, etc.

Geographical Distribution.   In closely situated islands, plants and animals are much alike; in more distantly separated islands, similarities are fewer. In regions shut off by barriers such as high mountains, wide expanses of water, etc., there are usually found certain species peculiar to that region but usually related to those of adjacent regions; it is believed that isolation leads to the development of certain types of organisms peculiar to the isolated areas.

Genetics and Selective Breeding.   These show that new types of plants and animals can be developed. The appearance of mutations is evidence of changes in the nature of living organisms.

Artificially Induced Changes.   By means of X-rays and other treatments changes can occur which are further evidence for the fact that living organisms are capable of evolution.

THEORIES ATTEMPTING TO EXPLAIN EVOLUTION

Lamarck's Theory of Use and Disuse (1809).   When an organism uses an organ, that organ is increased in size and functional ability. Failure to use a structure results in its retrogression or disappearance. According to Lamarck, changes in environment cause certain organs to be used, others not to be used, and thus cause the building-up or atrophy of organs. These changes in organs, according to Lamarck, are inherited. There is no evidence for the inheritance of such "acquired characters," so this theory is considered as unimportant at present.

De Vries' Mutation Theory (1901). Mutations, or sudden unpredictable, heritable changes which occur in certain organisms, are thought by De Vries to be the chief method by means of which new types of organisms develop. Mutations are known to occur in many, perhaps all, organisms, and they are probably responsible for the main developments in evolution.

Darwin's Theory of Natural Selection (1859). According to Darwin:

1. Organisms produce more offspring than can ordinarily survive.

2. These large numbers of offspring result in a competition for food, or a struggle for existence.

3. Individuals or species vary in their fitness to compete in this struggle. Those organisms which are best fitted for this struggle survive; those poorly fitted are eliminated in the struggle.

4. The offspring of parents best fitted to survive inherit the favorable characteristics of their parents.

Some of Darwin's statements are correct, but objections have been raised to portions of this theory so that at present the theory of natural selection is only partly accepted. Modern theories of evolution emphasize the importance of mutations and the probability of natural selection among these mutations. Hybridization has also been regarded as a possible explanation of certain phases of evolution.

THE COURSE OF EVOLUTION IN PLANTS

The major evolutionary changes which have occurred in the plant kingdom are:

1. The development of nuclei and other complex organelles, such as chloroplasts and mitochondria in cells.

2. Evolution from unicellular organisms to colonies to multicellular organisms, and from colonies of similar cells to colonies with cellular diversity.

3. Evolution of reproduction from asexual to isogamy to heterogamy. Development of alternation of generations.

4. Evolution of one-celled to many-celled sex organs.

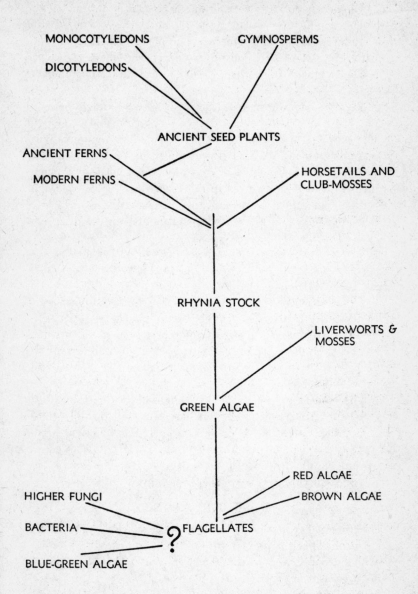

Fig. 24/1.   A suggested scheme of the evolution of the plant kingdom.

5. Change from water to land habitat.

6. Development of vascular and strengthening systems.

7. Progressive development of sporophyte generation, retrogressive development of gametophyte generation.

8. Development of flowers, seeds, and fruits which enclose seeds.

9. Development of highly efficient vessels in angiospermous wood.

The supposed evolutionary tree of the major groups of living plants is shown in Fig. 24/1. It will be noted that the course of evolution is not represented by a straight line, but that a much-branched diagram is required to represent the supposed paths of evolution. The salient features of this diagram are:

1. The flagellates are possibly ancestors of both the plant and animal kingdom.

2. The red, brown, and blue-green algae are considered as terminal groups (groups which have not led on to the evolution of higher forms of similar type).

3. The position of the fungi is uncertain. Some botanists consider them as possible descendants of the flagellates, others believe that they have evolved from different groups of higher algae.

4. The green algae have probably given rise to the Bryophyta and, by way of the extinct *Rhynia* stock, to the club-mosses, horsetails, and ferns.

5. The modern gymnosperms and angiosperms are thought to have evolved from a group of fernlike ancestors, possibly those which produced seedlike structures in their reproduction. Modern gymnosperms and angiosperms may thus be regarded as somewhat parallel lines which have evolved from a common ancestor or from closely-related ancestors.

FOSSIL PLANTS

The principal geological eras and the characteristic plant fossils found in rocks of those eras are:

Archaeozoic Era. Over one billion years ago. No signs of living organisms, except possibly bacteria.

Proterozoic Era. About one-half billion to one billion years ago. Some fossil evidence of the presence of blue-green algae and bacteria.

Paleozoic Era.   About 200 to 500 million years ago.

1. CAMBRIAN PERIOD.   500 to 600 million years ago. Fossil algae.

2. ORDOVICIAN PERIOD.   450 to 500 million years ago. Fossil algae.

3. SILURIAN PERIOD.   400 to 450 million years ago. Fossil algae.

4. DEVONIAN PERIOD.   350 to 400 million years ago. Low-growing, slightly-branched leafless and rootless *Psilophytes* (*Rhynia* and *Hornea*). Thought to be first land plants; bodies thalloid, but had vascular systems of a primitive type; considered as possible ancestors of ferns. In late Devonian rocks, many fossils of ferns, horsetails, and club mosses are found.

5. CARBONIFEROUS PERIOD.   280 to 350 million years ago. Dominant plants were ferns and their allies: ancient horsetails, club mosses, seed ferns, many growing as large trees over 100 feet tall. Bryophyte fossils found here.

6. PERMIAN PERIOD.   180 to 280 million years ago. Beginning of extinction of many carboniferous ferns, club mosses, etc. Appearance of cycads, ginkgos, conifers.

Mesozoic Era. 60 to 180 million years ago.

1. TRIASSIC PERIOD.   Many ancient ferns and early gymnosperms became extinct. Ginkgo very common.

2. JURASSIC PERIOD.   120 to 180 million years ago. Beginning of development of modern types of ferns.

3. CRETACEOUS PERIOD.   60 to 120 million years ago. Decline of certain gymnosperms. First appearance of angiosperms— willows, elms, magnolias, figs, beech, etc.

Cenozoic Era.   Present time to 60 million years ago. The modern era in which the angiosperms have become the dominant plants. Extinction of many species by glaciation. Large increase in number of herbaceous species.

Modern geologists do not agree on the delimitations of the Proterozoic and Archeozoic Eras. A possible compromise is to consider all time before the Paleozoic simply as pre-Cambrian.

Chapter 25

Plant Ecology and Geography

Plant ecology is the branch of botany which deals with the effects of environmental factors upon plant growth and distribution, and conversely, with the effects of plants upon the earth, air, water, and other organisms. *Plant geography* deals with the geographical distribution of various types of plants upon the earth's surface. *Autecology* is concerned with the individual organism and its relation to its environment; *synecology* is concerned with the total complex of inter-relations between organisms and between them and their environment.

ECOLOGICAL FACTORS

Environmental factors which influence the growth and distribution of plants are:

Climatic Factors (Factors of the Atmosphere).

TEMPERATURE. Temperature influences the rate at which the physiological activities of plants occur and thus influences plant growth and distribution.

LIGHT. Light influences plant growth through its effects on photosynthesis, transpiration, direction of growth, heating effect, flower production, enzyme action, etc.

CARBON DIOXIDE CONCENTRATION. Carbon dioxide concentration is the limiting factor in photosynthesis and thus changes in it affect the rate of photosynthesis. Excess CO_2 often inhibits growth.

OTHER ATMOSPHERIC GASES. Gases from smelters, furnaces, etc., are often injurious to vegetation.

WIND. Affects transpiration and also exerts mechanical effect upon direction of growth, form of plant, etc.

ATMOSPHERIC HUMIDITY AND PRECIPITATION.   Water vapor content of air, rain, snow, etc., are important factors influencing plant growth and distribution.

Edaphic Factors (Soil Factors).

AVAILABLE SOIL WATER.   One of the most important edaphic factors. Plants are often classified on the basis of their adaptation to water supply as follows:

Hydrophytes—plants which inhabit water or very wet soil: cattails, water lilies, pond weeds, etc. Usually weak-stemmed, with numerous air spaces, often very fine, thin leaves, and little or no cutin.

Xerophytes—plants which inhabit arid or semi-arid regions: cacti, sagebrush, Russian thistle, etc. Leaves absent or much reduced in size, usually with heavy layers of cutin, well-developed water-storage tissues, stomata often reduced in number and sunken in pits, etc.

Mesophytes—plants inhabiting regions with moderate water supply: common field and forest plants, such as roses, elms, maples, trillium, bluegrass, violets, oaks, etc.

Halophytes—plants which inhabit soils with high salt content and which can absorb water only with difficulty, because of high solute concentration of soil solution. Halophytes often resemble xerophytes structurally because of difficulty of absorbing water: salt bush, greasewood, etc.

SOIL TEMPERATURE.   Affects rate of root growth, respiration, water absorption, etc.

AIR IN THE SOIL.   Roots normally require oxygen for respiration and thus a deficiency of air affects root growth. Waterlogged soils have little air and plants growing in such soils are usually badly stunted.

PHYSICO-CHEMICAL NATURE OF SOIL.   Acidity or alkalinity of soil influence rate of absorption, nature of materials absorbed, and other activities of roots. Presence or absence of essential mineral elements influences rate of food synthesis and other anabolic processes in plants, etc. Physico-electrical properties of soil particles affect absorption, drainage, etc.

Biotic Factors (Effects of Other Living Organisms).

GRAZING BY ANIMALS.   Grazing removes food-making organs,

causes stunting of growth, etc. Also, mechanical effect of trampling.

SOIL ORGANISMS. Bacteria, fungi, algae, protozoa, worms in soil are important agents in increasing or decreasing soil fertility, in altering physical properties of soils, in attacking roots of higher plants, etc.

INSECTS. Insects which eat or otherwise injure plants or their parts or which function in pollination are important in growth, seed production, etc.

PARASITIC FUNGI. Disease-producing fungi may stunt growth, form abnormal growths, and generally decrease populations of susceptible plants, often to the point of extinction in certain regions.

COMPETITION. Struggle among various species of plants for water, soil salts, light, air, etc., influences distribution of plants.

SYMBIOTIC RELATIONSHIPS. Certain plants may cooperate for mutual benefit, as, algae and fungi in lichens, nodule bacteria on legume roots, etc.

ASPECTS OF ECOLOGICAL BEHAVIOR

Plant Communities. A group of plants which live together under the same set of environmental conditions is termed a *plant community*. Many kinds of plants may be present in a community, but usually one to three kinds of plants occur in greatest abundance, as often indicated by such names as oak-hickory forest, beech-maple-hemlock forest, etc. There are many types of plant associations, such as pond communities, xerophytic communities, mesophytic communities, swamp communities, etc.

Plant Succession. Plant succession is a series of changes in the plants of a given region as a result of disturbances in climatic, biotic, or topographic factors. In a given region, the first community which becomes established is termed a *pioneer community*. As conditions change, such communities may be followed by *intermediate communities*, which as a result of the activities of the pioneer organisms are better adjusted to the newer environmental factors than are the pioneer communities. The intermediate communities are followed usually with further changes in external factors by a *climax community*, the highest type which can be supported by the obtaining set of external factors and the type which tends to perpetuate itself year after year under similar conditions and does not render the environment unsuitable for its own offspring (e.g., grasslands of Kansas, eastern U. S. deciduous forest, evergreen forest of Rocky Mts.).

A common type of succession is that from pond community to swamp community to wet meadow to wet thicket or forest to a mesophytic climax forest community. If a climax community is destroyed by fire or some other factor, eventually the community will normally reappear by passing through the same series of changes.

Plant Invasion.   This is the tendency for a species or an association of species to extend its regions of occupancy and thus to invade new areas. Invasion is a common phenomenon in plant successions.

PLANT GEOGRAPHY

Whereas ecology is concerned chiefly with present factors and their relation to plant distribution, plant geography is concerned, in addition, with factors of past ages which have been important in determining the geographical distribution of various kinds of plants on the earth's surface. The plant geographer relies upon such geological phenomena as the relation of the continental land masses to each other, the emergence and subsidence of land masses, glaciation, ocean currents, changes in the earth's temperature, etc., for the explanation of the facts of plant distribution. Important factors in geographical distribution, in addition to the climatic, edaphic, and biotic factors described in the preceding section, are *barriers* (oceans, high mountain ranges, deserts, etc.) which prevent or discourage the migration of certain plant species and *highways* (mountain passes, rivers, etc.) which facilitate the migration of species. Often a barrier for one species may be a highway for another; oceans are barriers to most species of plants, but are often highways to species with resistant, floating seeds or fruits.

Various plant geographers have divided the earth's surface into plant geographical areas on the basis of vegetational types, geologic history, climatic features, etc. It is impossible to describe in this outline all of these areas; only the plant geographical areas of North America will be treated, as follows (Fig. 25/1):

Fig. 25/1. Map of vegetation regions of North America.

Tundra. Occupies the northern edges of North America, from Alaska to Labrador. Cold climate, with frozen soil, except for upper few inches in summer. Characteristic plants: lichens, mosses, grasses, sedges, other herbs, and a few shrubs.

Northern Evergreen Forest. Across continent from Pacific to Atlantic, south from tundra to Vermont, Great Lakes, and thence northwest through Canada and Alaska. Characteristic plants: chiefly conifers, such as black spruce, white spruce, hemlock, white pine, jack pine, balsam fir, etc.; some deciduous trees, such as aspens and birches, chiefly in cut or burned areas.

Deciduous Forest. From northern evergreen forest south along Appalachians to Texas and Louisiana, and westward from New York to Oklahoma, Wisconsin, and Minnesota. Characteristic plants: white oak, black oak, hickories, maples, chestnut, elms, walnut, ash, birch, tulip tree, and some conifers, such as short-leaf pine, hemlock, juniper, etc.

Southern Evergreen Forest. Coastal plain region, from Virginia to Texas, except Southern Florida. Characteristic plants: live oak, bald cypress, long-leaf pine, magnolia, gums, some short-leaf pine, etc.

Tropical Forests. Southern Florida, Central America, West Indies, Mexican coasts, etc. Characteristic plants: palms, *lianas* (woody climbers), orchids, mangroves, etc.

Grasslands. Texas to Manitoba, northwest into Canada, and west to the Rocky Mts., across Kansas, Nebraska, the Dakotas, etc. Characteristic plants: many species of grasses, asters, golden rods, sunflowers, etc. Relatively few trees.

Desert Regions. South from western Idaho and eastern Oregon through most of the region (The Great Basin) between the mountains of Colorado and Wyoming to the Sierra Nevadas of California and south Arizona, New Mexico, Texas, and Northern Mexico and Lower California. Characteristic plants: sagebrush, cacti, creosote bush, bunch grasses, rabbit brush, etc.

Rocky Mt. Forest. Southward from southeast Alaska and northwest Canada through Idaho, Montana, and the regions lying between the Grasslands and the Great Basin, extending south through the mountains of New Mexico into Mexico. Characteristic plants: western yellow pine, lodgepole pine, firs, western larch, some Douglas fir.

Pacific Coast Forest. The slopes of the coastal mountains, from southern Alaska into southern California. Characteristic plants: Sitka spruce, western hemlock, Douglas fir, western white pine, redwoods, western white cedar, arbor vitae, western yellow pine.

Chapter 26

The Importance of Plants in Human Life

The principal groups of products which are derived from plants and which are necessary, useful, or beneficial in human life are:

Fossil Fuels. Coal (the carbonized remains of compressed carboniferous forests), peat (the partially carbonized remains of recent sphagnum deposits), and petroleum (hydrocarbons from ancient plants, mainly diatoms), are the main heat and energy sources for present civilization.

Fibers. These include cotton, common hemp, sisal hemp, Manila hemp, jute, kapok, flax, ramie, bowstring hemp, coconut fiber, palm fibers, and others. Some fibers are derived from stems (linen, common hemp, jute, ramie); some from leaves (sisal hemp, Manila hemp, bowstring hemp, palm fibers); and some from fruits and seeds (cotton, kapok, coconut fiber). Principal products of fibers are: textiles, brushes, hats, baskets, chairseats, matting, upholstery and bedding stuffing, packing materials, string, twine, rope, nets, and caulking materials.

Wood Products. Fuel, lumber for construction and furniture making, fence posts, mine timbers, poles, pilings, pulpwood, cooperage, railroad ties, shingles, laths, veneers, plywoods, etc. Hardwoods are obtained from angiosperms, softwoods from gymnosperms.

Wood Derivatives. These substances are derived by chemical and physical alteration of wood. They include: charcoal, wood alcohol, acetate of lime, wood tar, wood gases, turpentine, oils, etc., all derived by wood distillation or wood extraction. Other derivatives of wood are: tannins (used in tanning hides), dyes, resins, essential (aromatic) oils, cellulose products (rayon, paper, pyroxylin, celluloid, artificial fabrics, cellulose varnishes), and drugs.

Tanning Materials. Used in tanning animal hides to convert them into leather, these materials are obtained from barks (hemlock, oak, mangrove, larch, Norway spruce, birch, willow, etc); from woods (chestnut, quebracho); from leaves (sumac); from roots (palmetto); from fruits (divi-divi).

Dyes and Pigments. These are obtained from woods (logwood, fustic, cutch, osage orange); from leaves (indigo, chlorophyll, henna); from roots and tubers (madder, turmeric, alkannin); from bark (black oak); from flowers (saffron); from seeds (annatto); from lichens (litmus). They are used for: dyeing textiles and leather, for paints, varnishes, paper, wood, and ink, and for adding color to beverages, medicines, and foods.

Gums. Gums are mucilaginous, water-soluble substances developing chiefly from the decomposition of cellulose and other complex carbohydrates. Principal plant gums are: gum arabic, gum tragacanth, gum karaya, peach gum, and cherry gum. They are used chiefly in mucilages and glues, paper sizings, calico printing, painting glazes, as stiffeners in ice creams, meringues, and other confections, as an adhesive agent in pills, as a stabilizing agent for insoluble powders in liquid medicines, and as soothing agent in medicines.

Resins. Resins are water-insoluble, oxidation products of various essential oils, formed in wood and bark of many woody plants. Used in manufacture of varnishes, lacquers, paper sizings, sealing wax, incense, perfumes, paints, linoleum, ink, and in medicine as healing agents and antiseptics in salves and ointments.

Latex Products. These are derived from the milky juice of stems and roots of various species of plants. Upon exposure to air or upon chemical treatment, latex solidifies into elastic or resilient substances, most important of which are rubber, chicle, gutta-percha, balata. Chicle is used in manufacture of chewing gum. Gutta-percha and balatta are used for insulating submarine cables, and for the manufacture of machine-belting, golf balls, telephone receiver cases and mouthpieces, surgical splints, dental fillings of temporary nature, etc.

Fats and Oils. *Fatty oils* and *waxes* are true fats and oils and their derivatives, and are obtained chiefly from seeds, or less frequently, from fruits and leaves. *Drying oils* harden into elastic films on exposure to air; they include linseed oil (from flax seeds), tung oil, walnut oil, perilla oil, and others. *Semi-drying oils* form soft, solid films only after longer exposure to air; they include cotton-seed oil, soybean oil, corn oil, sesame oil. *Non-drying oils* do not form solid films on exposure to air; include olive oil, castor oil, peanut oil. *Vegetable fats* are solid or semi-solid at ordinary temperatures; include coconut oil, palm oil, cocoa butter, nutmeg butter. Chief uses of oils and fats: foods, soaps, candles, linoleum, paints, varnishes, furniture and leather polishes, oil papers, waterproof fabrics, inks, artificial leather, illuminants, oilcloth, cooking oils, putty, laxatives, lubricants, cosmetics, massaging oils. *Waxes* are similar to fats chemically, but are harder; they are derived chiefly from leaves (carnauba wax) and fruits (wax myrtle) and are used in the manufacture of candles, wax varnishes, phonograph records, floor- and shoe-polishes, etc.

Smoking and Chewing Materials. Tobacco is the only important smoking material. Chewing products, in addition to chicle, include betel nuts, cola nuts, and coca leaves. Opium and marijuana are smoked illegally for their narcotic effects.

Drugs. Used for many purposes in medicine, some drugs are aconite, ginseng, ipecac, licorice, podophyllum, cascara, quinine, morphine, belladonna, ephedrine, digitalis, witch hazel, senna, croton oil, agar, ergot, and many others.

Beverages. Used for pleasant flavors and often mildly stimulating effects, they include coffee, cocoa (cacao), tea, maté, and cola. The stimulating qualities of such beverages are chiefly attributable to alkaloids, such as caffeine.

Foods. Foods are used by man for his nourishment and for that of his domesticated animals.

GRAINS. The basic food plants of man. Include wheat, corn (maize), rice, barley, rye, oats, sorghum, kafir, milo, millets, etc. The great civilizations of the earth have been based upon rice (Oriental), wheat (European), and corn (American).

LEGUMES. Peas, beans, soybeans, cowpeas, peanuts, lentils.

NUTS. Brazil nuts, coconuts, cashew nuts, pecans, walnuts, hickory nuts, hazelnuts, almonds, etc.

VEGETABLES. Vegetables come from various parts of plants: from roots—beets, carrots, radishes, turnips, parsnips, sweet potatoes, yams, cassava; from underground stems—Irish potatoes, onions, taros; from shoots (leaves and stems)—asparagus, arti-

chokes, cabbage, Brussels sprouts, broccoli, kohlrabi, celery, lettuce, rhubarb, spinach; from flowers or fruits (botanical fruits)—
avocado, breadfruit, eggplant, cucumber, pumpkin, squash, tomato, okra, cauliflower.

TEMPERATE ZONE FRUITS.   Apple, pear, apricot, cherry, quince,
peach, plum, watermelon, muskmelon, grape, blackberry, raspberry, blueberry, cranberry, gooseberry, currant, strawberry.

TROPICAL AND SUBTROPICAL FRUITS.   Orange, lemon, grapefruit,
tangerine, citron, lime, banana, plantain, custard apple, pineapple,
mango, date, fig, guava, mangosteen, papaya, pomegranate,
sapodilla.

SPICES AND OTHER AROMATIC SUBSTANCES.   These are used for
flavoring foods, beverages, medicines, tobacco, and confections and
for scenting cosmetics, soaps, and incense. Also used in medicine
as carminatives and antiseptics. Among the most important plants
in this class are: ginger, angelica, anise, cinnamon, sassafras,
cloves, allspice, peppers, juniper, vanilla, dill, caraway, cardamom,
mustard, nutmeg, mace, tonka beans, sage, peppermint, spearmint,
thyme, bay, wintergreen, parsley, paprika, rose, citronella, lemon
grass, lavender, jasmine, camphor, geranium, eucalyptus, orange,
lemon, lime, rosemary, and sandalwood.

All the above economically important plant products are derived principally from seed plants (Spermatophyta). Of the lower
groups plants, living Pteridophyta and Bryophyta are of virtually
no economic importance. The Pteridophyta of past ages, and some
of their near-relatives, were largely responsible for the formation
of coal deposits.

Of the lower divisions of the plant kingdom, the Thallophyta
are of greatest importance to man. The algae, or chlorophyllous
Thallophyta, are important chiefly in the following ways: food
for human beings, food for fish and other animals of ponds, lakes,
rivers, and oceans, food for protozoa, worms, and other soil animals; sources of fertilizer for soils, of medicinal iodine, of mucilaginous substances such as agar-agar and others used in culture
media for bacteria and other fungi; and for stiffening confections,
ice creams, shaving creams, mucilages, etc. Algae are also important to man in a harmful or deleterious manner in that they
pollute water supplies in which they produce offensive odors and
flavors; in some cases, poisoning of cattle and other domestic ani-

mals has resulted from drinking algae-polluted water. Algae in the sea are largely responsible for the balance of oxygen and carbon dioxide in the earth's atmosphere.

The fungi (in the broad sense of the word) are exceedingly important economically. Among the beneficial or advantageous uses of fungi are: food for man and other animals, production of vitamins, penicillin, ergotine, and other valuable drugs, manufacture of cheese and other dairy products, production of alcoholic beverages and of industrial chemicals such as alcohols, acetone, enzymes, organic acids, etc., sanitation of the earth's surface and maintenance of soil fertility by decomposing the dead bodies and waste products of plants and animals, the production of sauerkraut, vinegar, and other important foods, the retting of flax and other fibers, the removal of pulp from coffee and cocoa beans, the curing of vanilla pods, etc. Fungi are also important to man in a harmful or deleterious manner: they cause diseases of man (tuberculosis, pneumonia, leprosy, typhoid fever, lockjaw, various skin diseases such as athlete's foot, ring-worm, etc.); they cause destructive diseases of economically-valuable higher plants (wheat rust, cornsmut, apple scab, chestnut blight, wilt diseases of squash, tomatoes, cucumbers, potatoes, etc., potato blight, mildews of grape, peach brown rot, etc.); they cause the spoilage of foods (souring of milk, spoilage of meats, canned foods, fruits in transit); they cause the decay of wood, rotting of leather and fabrics and other objects rich in organic matter; and they cause many diseases of man's domesticated animals (cholera, pneumonia, anthrax, glanders, tuberculosis, etc.).

Plants also influence human life in less direct ways than those mentioned above:

1. Plants provide food and shelter for wild animal life.

2. Plants bind the soil and thus reduce or prevent erosion by wind and water.

3. Weeds reduce the yields of field and garden crops and disfigure lawns and ornamental gardens. Weeds compete with desirable plants for light, water, soil nutrients, and space, and thus reduce the growth and yield of the latter. Weeds may also mechanically injure garden and field plants by growing upon them and breaking off their stems, and they often harbor insect pests and pathogenic fungi which attack crop plants.

4. Poisonous plants are sometimes eaten by sheep, cattle, and other domesticated animals. They may cause violent illness, reduction in milk yield, abortion, poisoning of milk, and death of the animals.

5. Masses of vegetation influence the temperature and humidity of the atmosphere and thus may affect precipitation and other climatic phenomena.

6. Wild and cultivated plants contribute to man's esthetic life.

7. Many theoretical discoveries of prime biological importance have been made by the use of plants as experimental tools, as in genetics, metabolism, etc.

The exploitation of economically valuable plants has had many far-reaching social, political, and historical effects upon human life. For example:

1. The social and medical problem of drug addiction is based largely upon morphine, cocaine, and marijuana—all derived from plants. During World War II, Japanese occupation forces in China encouraged the cultivation and traffic in poppies, since they knew that opium, derived from these poppies, would weaken the physical and mental resistance of the Chinese who made use of them.

2. The social problem of slavery was intimately associated in earlier times with the exploitation of cotton in the southern United States, of sugar cane in the West Indies, of rubber in Africa.

3. The potato famine in Ireland in 1845, which resulted from a serious outbreak of the potato blight disease, caused by a fungus, was responsible for widespread misery and death in that country and for the emigration of thousands of Irish people to the United States.

4. History's most famous mutiny, that against Captain Bligh, by the crew of the "Bounty," developed on a cruise which had as its principal object the introduction of an important food plant, breadfruit, from islands of the South Pacific to the English colonies of the West Indies.

5. The voyages of Columbus (1492) and of Vasco de Gama (1497) in an attempt to find a water route to India resulted in large part from a desire to exploit and control the trade in spices, sugar, and other valuable plant products of the Orient.

6. The expansion of Japan and Germany in Asia and Central Europe respectively was partly a result of overpopulation and the need for additional agricultural land to produce food. This was one of the major causes of World War II.

7. The crippling effect of the rubber shortage in 1942–1946 in the United States arose from Japan's capture of the world's principal rubber plantations, those in Malaya, Indo-China, and Sumatra, in early 1942.

8. Overproduction of crops has on many occasions (as, for example, in the United States in the early 1930's) resulted in unprofitable prices for farmers and gardeners and has led to such drastic economic regulations as destruction of crops, removal of acreage from cultivation, price subsidies, special taxes and tariffs, etc.

9. The miserable plight of migratory farm workers and share-croppers at various times has been related to agricultural maladjustments.

10. The failure of food production to keep pace with the increase in human population is posing a serious threat to the stability of nations.

Selected Bibliography
and Index

Selected Bibliography

Baerg, H. *How to Know the Western Trees.* ("Pictured-Key Nature Series.") W.C. Brown, Dubuque, Iowa. 1955.

Baker, J.J.W. and Allen, G.E. *Matter, Energy, and Life.* ("Principles of Biology Series.") Addison-Wesley, Reading, Mass. 1965.

Barry, J.M. *Molecular Biology: Genes and the Chemical Control of Living Cells.* ("Concepts of Modern Biology Series.") Prentice-Hall, Englewood Cliffs, N.J. 1964.

Bennett, T.P. and Frieden, E. *Modern Topics in Biochemistry: Structure and Function of Biological Molecules.* Macmillan, New York. 1966.

Billings, W.D. *Plants and the Ecosystem.* ("Fundamentals of Botany Series.") Wadsworth, Belmont, Calif. 1964.

Bold, H.C. *The Plant Kingdom.* ("Foundations of Modern Biology Series.") 2nd ed., Prentice-Hall, Englewood Cliffs, N.J. 1964.

Brook, A. *The Living Plant.* Aldine, Chicago, Ill. 1967.

Conard, H.S. *How to Know the Mosses and Liverworts.* ("Pictured-Key Nature Series.") W.C. Brown, Dubuque, Iowa. 1956.

Cook, Stanton A. *Reproduction, Heredity, and Sexuality.* ("Fundamentals of Botany Series.") Wadsworth, Belmont, Calif. 1964.

Dawson, E.Y. *Marine Botany: An Introduction.* Holt, Rinehart and Winston, New York, 1966.

———. *How to Know the Seaweeds.* ("Pictured-Key Nature Series.") W.C. Brown, Dubuque, Iowa. 1956.

Delevoryas, T. *Plant Diversification.* ("Modern Biology Series.") Holt, Rinehart and Winston, New York, 1966.

Doyle, W.T. *Nonvascular Plants: Form and Function.* ("Fundamentals of Botany Series.") Wadsworth, Belmont, Calif. 1964

Galston, A.W. *The Life of the Green Plant.* ("Foundations of Modern Biology Series.") 2nd ed., Prentice-Hall, Englewood Cliffs, N.J. 1964.

Jaques, H.E. *How to Know the Trees.* ("Pictured-Key Nature Series.") W.C. Brown, Dubuque, Iowa. 1946.

Jensen, W.A. *The Plant Cell*. ("Fundamentals of Botany Series.") Wadsworth, Belmont, Calif. 1964.

———, and Kavaljian, L.G. *Plant Biology Today: Advances and Challenges*. Wadsworth, Belmont, Calif. 1966.

Johnson, W.H. and Steer, W.C., eds. *This is Life: Essays in Modern Biology*. Holt, Rinehart and Winston, New York. 1962.

Knobloch, I.W., ed. *Selected Botanical Papers*. ("Biological Science Series.") Prentice-Hall, Englewood Cliffs, N.J. 1963.

Lee, A.E. and Heimsch, C. *Development and Structure of Plants: A Photographic Study*. Holt, Rinehart and Winston, New York. 1962.

Levine, R.P. *Genetics*. ("Modern Biology Series.") Holt, Rinehart and Winston, New York. 1962.

Loewy, A.G. and Siekevitz, P. *Cell Structure and Function*. ("Modern Biology Series.") Holt, Rinehart and Winston, New York. 1963.

McElroy, W.D. *Cellular Physiology and Biochemistry*. ("Foundations of Modern Biology Series.") 2nd ed., Prentice-Hall, Englewood Cliffs, N.J. 1964.

Odum, E.P. *Ecology*. ("Modern Biology Series.") Holt, Rinehart and Winston, New York. 1963.

Pohl, R.W. *How to Know the Grasses*. ("Pictured-Key Nature Series.") W.C. Brown, Dubuque, Iowa. 1954.

Prescott, G.W. *How to Know the Freshwater Algae*. ("Pictured-Key Nature Series.") W.C. Brown, Dubuque, Iowa. 1954.

Ray, P.M. *The Living Plant*. ("Modern Biology Series.") Holt, Rinehart and Winston, New York. 1963.

Salisbury, F.B. and Parke, R.V. *Vascular Plants: Form and Function*. ("Fundamentals of Botany Series.") Wadsworth, Belmont, Calif. 1964.

Solbrig, O.T. *Evolution and Systematics*. ("Current Concepts in Biology Series.") Macmillan, New York. 1966.

Steward, F.C. *Plants at Work*. ("Principles of Biology Series.") Addison-Wesley, Reading, Mass. 1964.

———, Krikorian, A.D. and Holsten, R.D. *About Plants: Topics in Plant Biology*. ("Principles of Biology Series.") Addison-Wesley, Reading, Mass. 1966.

Swanson, C.P. *The Cell*. ("Foundation of Modern Biology Series.") 2nd ed., Prentice-Hall, Englewood Cliffs, N.J. 1964.

Went, F.W. *et al. The Plants*. ("*Life* Nature Library.") Time Inc., New York. 1963.

White, E.H. *Chemical Background for the Biological Sciences*. Prentice-Hall, Englewood Cliffs, N.J. 1964.

Index

Girdling, 60, 62
Gladiolus, 46
Glucose, 73 *ff.*
Glycerol, 80
Glycolysis, 85
Golden-brown algae, 153
Golgi bodies, 19, 94
Gooseberry, 166, 209
Grafting, 62
Grain, 131, 133
Grana, 18, 70, 71
Grape, 59, 64, 129, 162, 209
Grapefruit, 209
Grass, 8, 191
Grasslands, 205
Gravel, 32
Gravity, 113
Greasewood, 201
Greeks, 1
Green algae, 151
Green molds, 164
Ground pine, 181
Growth, 8, 91 *ff.*, 101 *ff.*, 112 *ff.*
GTP (*see* guanosine triphosphate)
Guanine, 102
Guano, 79
Guanosine triphosphate, 108, 109
Guard cells, 21, 67
Guava, 209
Gullet, 150
Gums, 54, 61, 207
Guttation, 77
Gymnosperms, 49, 51, 54, 187

Hairs, 45
Halophyte, 201
Haploid chromosome number, 141
Haploid phase, 97, 98
Hardwoods, 52, 206
Hawthorn, 166
Hazelnut, 208
Head, of composite, 124
Heartwood, 54
Heat, 85
Hemlock, 204

Hemp, 49, 206
Hepaticae, 173
Herbaceous plants, 47
Herbaceous stems, 54
Herbals, 1
Heterochromatin, 94
Heteroecious rust, 166
Heterogamy, 151
Heterospory, 100, 178, 186
Heterothallism, 163
Heterotrophic plants, 83, 148
Hickory, 205, 208
High-energy bond, 72
Hill reaction, 71
Hilum, 132
Histones, 101
Hofmeister, 13
Hog cholera, 159
Holdfast, 151
Homeostasis, 111
Homologue (chromosome), 96
Homospory, 100, 178
Homozygous condition, 138
Honey locust, 46
Honeysuckle, 123
Hooke, Robert, 12
Hormones, 79, 120
Hornwort, 175
Horseradish, 161
Horsetail, 181
Horticulture, 2
Humidity, 76, 201
Humus, 33
Hybrids, 63
Hydathode, 77
Hydrogen acceptor, 71
Hydrogen bonds, 106
Hydrolysis, 81
Hydrophyte, 201
Hydroponics, 79
Hydrostatic pressure, 27
Hydrotropism, 119
Hygroscopic water, 32
Hypha, 161
Hypocotyl, 35, 42, 133
Hypogynous flowers, 122
Hyponasty, 119